Disequilibrium Economics

Tönu Puu

Disequilibrium Economics

Oligopoly, Trade, and Macrodynamics

Springer

Tõnu Puu
Centre for Regional Science (CERUM)
Umeå University
Umeå, Sweden

ISBN 978-3-030-08986-3 ISBN 978-3-319-74415-5 (eBook)
https://doi.org/10.1007/978-3-319-74415-5

Softcover re-print of the Hardcover 1st edition 2018

Printed on acid-free paper

This Springer imprint is published by the registered company Springer International Publishing AG part of Springer Nature.
The registered company address is: Gewerbestrasse 11, 6330 Cham, Switzerland

To the memory of Martin J Beckmann
1924–2017

Preface

This book is titled "disequilibrium economics" to mark contrast and distance to the equilibrium outlook that dominated economics since ages. That outlook was deeply rooted in tradition but came to dominate particularly since about 1950 when the existence proofs for general equilibrium met with final success.

One may evaluate this celebrated theory differently. From an epistemological point of view, the present author is not particularly impressed, though he understands that others may be thrilled with the artwork of the final proof, pending since the days Walras counted independent variables and equations.

The big problem with any dominant paradigm is that it suffocates alternative approaches, in this case promising dynamics from the period 1920 to 1950, forgotten or relegated to minor branches of science.

Competing outlooks, such as monopolistic competition, were deleted from the central body of theory, and, most remarkably, Keynesian macroeconomics was destroyed through attempts to "*derive*" it from "*microeconomic foundations*". As if physicists had attempted to *derive* the Maxwell-Boltzmann energy distribution in a dilute gas from Newton's mechanics.

As for dynamics, the study of nonlinear processes in the natural sciences (1970–1980) gave an impetus also to economics, though mainly thanks to physicists and mathematicians who looked for new applications in modelling social phenomena. Economists did not much care for it.

The inspiration for the present book comes from a rich tradition of the today largely neglected fields of dynamic oligopoly and macrodynamics, from the work by Augustin Cournot, Richard Goodwin, John Hicks, Harold Hotelling, Tord Palander, Paul Samuelson, Heinrich von Stackelberg, Abraham Wald, and others; yes, indeed, even Wald, who in the very same article that provided an important stepping stone towards the existence proof for general equilibrium, also proposed duopoly models with infinite numbers of equilibria. However, nobody reads that part of his long article any longer.

The present author does not feel entirely alone in his admiration for the period of somewhat unstructured renewal. Joseph Schumpeter in his *History of Economic Analysis* ironically scorches the "classical situations" in which scientists think they understand everything in the frame of one single supertheory which has an "*aspect of repose and finality, like a Greek temple that spreads its perfect lines against a cloudless sky*".

No doubt it was difficult for economists to plunge into dynamic modelling as they tried it with linear approximations. These have but two global scenarios—return to eternal rest, or explosion that blows up the model.

Economics is a natural mathematical science, unique among the social sciences, as it deals with measurable quantities and their relations.

However, by tradition focus is on local and qualitative descriptions through sketchy graphs, tables, or verbal statements of functions in terms of slope, convexity, and the like. In calculus format the assumptions are about signs of derivatives. With this goes that whenever a specific function is assumed, doubt is raised that it is but a very special numerical case. Economists never sensed that nondegenerate functions can represent families of topological equivalents.

This habit may be one reason why economists became so hooked to existence proofs and fixed point theorems; these methods provide a means to at all proceed from the sketchy qualitative assumptions. However, the price is an extremely scant supply of precise conclusions, and lots of theoretical constructs that result in no conclusions at all.

Indirectly the focus also explains the concentration on normative rather than positive theory. Economics truly overflows with various optimality rules concerning everything, from benefit cost rules for large-scale infrastructure down to when to cut a tree or sell a bottle of wine. Unfortunately, the applicability is often too theoretical and vague for actual use.

Thus economics has little to offer in terms of nonlinear functions that hold over a substantial portion of phase space and hence are useful for global modelling. True, there are no laws of nature in economics, but one has to make the best of this situation and use whatever there exists of credible functions.

The field of nonlinear dynamics was left wide open for physicists and mathematicians who wanted to try their hands on economics modelling. Some even specialized in such and eventually called themselves economists.

The principles of oligopoly, invented by a mathematician, or business cycle theory, formed in analogy to mechanical oscillators, of course, are easy to grasp in a moment, and these were easy to grab for analysis, though there remains a slight danger with modellers who lack the intuition for economic phenomena that researchers of that discipline after all get through years of practice, priming and indoctrination.

The present author who is a pure economist by training, but who also learned some mathematics, has been working with nonlinear dynamics for about 30 years, and much of the material has already been published. New is that a variety of models from micro- and macroeconomics are collected in the uniform perspective of disequilibrium.

As for presentation, the author feels that pictures are essential for developing intuition, so he was delighted when the publisher agreed to illustrations in colour and he used considerable effort with the artwork, which is freshly cooked computer graphics.

As the author acquired some experience over 50 years of model making, but is not a mathematician himself, he appended a number of "*research stubs*"—research agendas described to some detail, but incompletely worked. He leaves them for free, and hopes they might inspire especially young interdisciplinary researchers for further analysis and study.

In the stubs, though not in the main text (with the sole exception of Hotelling duopoly), we also include a number of issues concerning geographical space. They deal with migration, interregional trade, and diffusion of business cycles. These models were originally cast in continuous settings, concerning space and time, and are now discretized for the first time.

We always have to make a choice whether to choose a continuous or a discrete representation as frame. As is well known many continuous time processes, such as logistic growth, are well behaved, whereas the discrete versions become manifestly unstable. This, of course, has to do with the intermittent stepwise adjustments that cause overshooting rather than smooth adaptation.

The author does not want to denounce continuous modelling in any way. It has been extremely useful through the history of science. However, it always remains an abstraction. Originally a numerical algorithm for some classical continuous model was an approximation that could "go wrong" if it did not converge. However, there is a suspicion that such approximations may even be closer to the facts of reality, as discrete data are all that we have in terms of information.

We must take advantage of a profound change of tools. The computer totally changed things. Not only do we know a lot more about discrete systems, but we are no longer tied to the after all limited supply of closed form solutions to differential equations and their interrelations which provide the basis of calculus. We can simply run any system without concern about whether it is related to already known relations or not.

This is well known to all as long as time is concerned. But in the case of space differential systems still dominate, and there has been remarkably little software developed for dealing with PDE systems. Therefore it seems important to start putting space on equal footing with time.

However, we have to recall a profound difference between time and space. Nonrelativistic time is related to causality and has a built-in forward direction. Space, even in one dimension, does not; every internal point is surrounded by points on both sides. It is for this reason lowest time derivatives in differential systems are first order, and space derivatives second order.

Spatial modelling in economics tends to simplify to the line. An important agenda is to take account of the full two dimensions of geographical space provided with some reasonable distance metric. It is not too difficult, and even simplified by the fact that more convenient metrics than the Euclidean may be appropriate. In economics we still have the huge advantage of only 2D compared to the 3D of classical mechanics.

Acknowledgements

The author has profited much from close collaboration with a few very good mathematicians. Though he, as an economist, assumes full responsibility for the model setup, the mathematical studies very much helped him to understand their working, and, even when seemingly pedantic, frequently disclosed snags and triggered reformulations of the models.

In earlier years the collaboration was with Professor Laura Gardini and her students at the University of Urbino, in particular Drs. Anna Agliari and Fabio Tramontana, and with Professor Gardini's close friend Dr. Iryna Sushko from the National Academy of Sciences of Ukraine in Kiev. In later years an equally yielding collaboration evolved with Professor Jose Cánovas at the Technical University of Cartagena and Dr. Anastasiia Panchuk, also from the NAS in Kiev, with whom the author spent many as pleasant as yielding working sessions as guests of Professor Cánovas in Cartagena. The author is deeply indebted to these scientists who have contributed much to the contents of this book.

Cartagena
October 2017

Tönu Puu

Contents

Part II Models

Part I
Introductory Matter

Chapter 1
Utility and Demand

Abstract Demand and supply are the fundament for almost all economic reasoning. The concepts are based on the idea that utility and scarcity jointly determine the value of commodities. The idea may go back as far as to Greek antiquity and Plato who wrote about almost everything. In this Chapter we attempt to collect utility functions which are useful for global dynamic systems. Particularly, we emphasize the usefulness of Lancaster's theory where it is not the marketed commodities that are entered in the utility functions, but their "properties" assumed to be measurable scores. This is a seldom used approach, which, however, gives us a unique possibility to model commodities that are close substitutes.

1.1 The Paradox of Value

Here we take the story from Schumpeter (1954).[1] Enigmatic is the *paradox of value*: "How can it be that almost useless commodities, such as gold and diamonds, are so valuable whereas fresh air to breath and water to drink are free?" Schumpeter refers to Bernardo Davanzati (1529–1606) who in his "*Lezioni delle monete*" (1588), when discussing the use of gold for coins, clearly raises this issue and identifies utility and scarcity as joint determinants for the high value of gold.[2]

Of course, scarcity and utility as explaining value are most useful if they are measurable quantities. No problems with scarcity, natural resources, such as pockets of ore, can be measured in terms of volume or weight. Also, if not open but

---

[1] This posthumous work is a real treasure, though very extensive with its 1260 pages in small print. It is also a bit fragmentary, which is explained by the fact that Schumpeter worked with it almost until his death, after which his collaborators continued using his notes. We will refer to this great work repeatedly, though reading from cover to cover takes a bit of patience. Fortunately, the book has an excellent index for topics and authors, so it is easy to find specific information.

[2] By the way, today's inhabitants of metropolis hardly agree that fresh air is a free commodity, and poor farmers in their daily fight for fresh water hardly agree about fresh water either. The latter is not historically new. European cities still around 1700 were deficient in water supply, which was replaced by an enormous consumption of alcohol.

T. Puu, *Disequilibrium Economics*, https://doi.org/10.1007/978-3-319-74415-5_1

needing extraction, labour and other resources needed are quantifiable. But, how about utility?

Enlightenment philosopher Jeremy Bentham in his "*Introduction to the Principles of Morals and Legislation*" (1789) considered measurable units of pleasure and pain that would put utility on equal footing with scarcity.[3]

As a companion to the idea of measurable utility the idea of decreasing marginal utility from consumption appeared. When satisfying needs, the first units of satisfaction were most gratifying, whereas after successive additions a point of saturation would be approached. This naturally applied to individual commodities, though it could also be used for total consumption. Decreasing marginal utility could be used as an "objective" argument for redistribution, as the rich would lose less happiness than the poor would gain.

The true heyday for "*marginalism*" came around 1870, with the double focus of Carl Menger[4] (1840–1921) and William Stanley Jevons[5] (1835–1882). The thrust was to consider that good economic decisions were taken on the basis, not of the total or the average value of some variable, but on the marginal value. For instance, it was appropriate to increase consumption as long as the utility of the last unit—the marginal utility exceeded the cost for acquiring this last unit.

Economists thus developed their own "infinitesimal" calculus long after it had been introduced by Descartes and Newton in mathematics two Centuries earlier.

1.1.1 Indifference

Utility as a measurable substance that could be counted or weighted was eventually abandoned. The central concepts became preference and indifference, and utility remained just a way of representing relations between choice alternatives of which one was preferred, or, in rare cases, both were equivalent. Utility became just a tag to fix on alternatives, where a higher utility indicated that this was preferred.

The important thing was that, whenever two alternatives were admissible, for instance, affordable given a budget to spend and the prices of commodities, the consumer would *always make the same choice*. Consistency in choice thus replaced the idea of utility as a measurable substance. This older idea was called *cardinal* utility, whereas utility as representing consistent preference ordering was named *ordinal*. The nomenclature of course originated from linguistics where cardinal counting numbers "one, two, three" were distinguished from ordinal "first, second, third" already present in Latin.

[3]One can still say hello to Bentham at University College in London, where, according to his will, his skeleton sits on a chair in a special wooden cabinet, in his own clothing stuffed with hay. Unfortunately, the preservation of his head went wrong, so it is replaced by wax.

[4]Menger (1871).

[5]Jevons (1871).

Such choice consistency was stripped of all value content, and represented just a kind of rationality. The ordinal theory of choice was provided with an axiomatic basis from mathematical logic for inequalities in terms of reflexivity, transitivity, and the like.

1.1.2 Altruism

Due to this focus on seeking pleasure and avoiding pain, and, later, making consistent and rational choices gave economics a bad reputation as a science focused on egotism. This is largely a misunderstanding, as a consistent preference system may well include altruism.

1.2 Utility and Demand Functions

For our purpose we want demand functions that: (1) work in a *global* setting, such as a dynamic orbit; (2) can be derived from some credible utility function.

1.2.1 Linear Demand

The first requisite is a problem in economics, because economic theory for ages has focused on *local* characteristics in static systems.[6]

The second requisite is seldom satisfied. A linear approximation is always handy, in economics as in physics. However, an approximation is something local as well.

Used in dynamic settings, it makes systems explode or going to eternal rest. Further, it can make variables (such as prices) become negative, which is nonsense in any economics context.

A linear demand function cannot be derived from any optimization problem. Sometimes it is argued that a linear demand function can be derived from a quadratic utility function, but this is not true. Probably the deception originates in the visual fact that expansion paths are straight lines, but these are not demand functions.

---

[6]For instance considering complements and substitutes among consumers' goods, one looks at the sign of the derivative of demand with respect to changes of price of another good. However, if we want to find a formula for a demand function that holds while circumstances change during a dynamic process, we find very little, especially if we want to portray groups of commodities as substitutes (goods that more or less satisfy the same needs) or complements (such that preferrably are consumed in combination).

To see this, assume utility U as a quadratic function of the consumption of two commodity quantities q_1 and q_2.[7] We take it as an upside down paraboloid[8]

$$U = 2 - (q_1 - 1)^2 - (q_2 - 1)^2 \tag{1.1}$$

A quadratic function necessarily has an extreme point. By the minus signs we make sure it is a maximum, not a minimum, and the point $(1, 1)$ grants that this maximum is located in the positive quadrant of the q_1, q_2-plane. Further, the added constant 2 grants that utility is positive in the region of interest for optimization, not logically necessary but pleasing for intuition.

Utility is considered to be maximized under a budget constraint, where the prices p_1, p_2 are taken as positive parameters, like the income budget y[9]

$$p_1 q_1 + p_2 q_2 = y \tag{1.2}$$

Solving this rather trivial constrained maximization problem, we get

$$q_1 = 1 - \frac{p_1 + p_2 - y}{p_1^2 + p_2^2} p_1 \tag{1.3}$$

$$q_2 = 1 - \frac{p_1 + p_2 - y}{p_1^2 + p_2^2} p_2$$

which notably are *far from linear*!

We also need a number of constraints. Of course, q_1, q_2 must be positive. Further, marginal utilities become negative whenever $q_1 > 1$ or $q_2 > 1$, so these cases must be ruled out.

---

[7]Economics has a generally accepted code of symbols which greatly simplifies understanding of a model: p for prices, q for quantities, k for capital, l for labour, r for capital rent, w for wage rate, i for interest rate, Y for income, I for investment, S for saving, etc. This is very convenient for fast recognition of a problem.

[8]This is totally symmetric. It can easily be given a more general shape through linear transformations $q_i = a_i x_i$

$$U = 2a_1 x_1 + 2a_2 x_2 - a_1^2 x_1^2 - a_2^2 x_2^2$$

though there is no point in complicating things. Economists some times have the bad habit of complicating things to make a general appearance, instead of stripping cases one wants to make to the essentials.

[9]It would be more appropriate to write

$$p_1 q_1 + p_2 q_2 \le y$$

i.e., expenditures may not *exceed* the available budget, but as long as consumption needs are not saturated we may use the equality. Though, particularly with the example of the paraboloid, it has a global maximum, beyond which greater expenditure decreases utility.

Combining, we get

$$0 < \frac{p_1 + p_2 - y}{p_1^2 + p_2^2} p_1 < 1 \tag{1.4}$$
$$0 < \frac{p_1 + p_2 - y}{p_1^2 + p_2^2} p_2 < 1$$

These conditions simplify to

$$p_1 + p_2 > y \tag{1.5}$$
$$\frac{(p_1 - p_2)^2}{p_1 + p_2} > y$$

By conclusion, from quadratic utility the demand functions are by no means linear, and they are subject to constraints, of which the second, in particular, rules out an extended region around the diagonal in p_1, p_2-space. This implies that the demand functions are particularly unsuitable for the study of close substitutes.

Further, linear demand functions cannot be derived from any budget constrained optimization as they do not satisfy the budget constraint. Suppose we have

$$q_1 = A_1 - B_1 p_1 + C_1 p_2 \tag{1.6}$$
$$q_2 = A_2 - B_2 p_2 + C_2 p_1$$

as suggested in some recent contributions. Substituting in the budget constraint we have

$$p_1 q_1 + p_2 q_2 = A_1 p_1 + A_2 p_2 + B_1 p_1^2 + B_2 p_2^2 - (C_1 + C_1)\, p_1 p_2 \tag{1.7}$$

which only by accident can equal the constant y.

By conclusion, linear demand functions do *not* result from quadratic utility functions, nor can they be outcomes of *any* constrained utility optimization.

Yet, if we bear in mind that they are approximations, they can be interesting to use, as they serve to teach how to handle constraints. Reasonable "linear" models in economics are therefore never linear, but piecewise linear, something that produces several interesting problems as we see in the sequel.

1.2.2 Cobb-Douglas Utility

A very useful shape for utility is the Cobb-Douglas function:

$$U = q_1^{\alpha_1} q_2^{\alpha_2} \tag{1.8}$$

It did not emerge as a function for utility, but rather as a production function, to organize production possibilities for firms from different and various branches.[10] Though, soon the function found its way to demand theory as a suitable utility function. Again, there is no point in emphasizing asymmetries, so we take

$$U = \sqrt{q_1 q_2} \tag{1.9}$$

In the application to production, as a firm's output is something measurable, the sum $\alpha_1 + \alpha_2 >$ or < 1 makes sense as watershed between increasing and decreasing returns to scale. In the context of ordinal utility this is not the case. Any monotonic transformation $U = q_1 q_2$ or even $U = \log q_1 + \log q_2$ represents the same preferences, even though to intuition the latter alternative may seem absurd as it returns negative utility for $q_1 q_2 < 1$.

Maximizing utility $U = \sqrt{q_1 q_2}$ under the budget constraint $p_1 q_1 + p_2 q_2 = y$ returns the very simple solution

$$q_1 = \frac{1}{2}\frac{y}{p_1} \tag{1.10}$$

$$q_2 = \frac{1}{2}\frac{y}{p_2}$$

Consumers spend half their income on each commodity.[11] Note that the demand for q_1 only depends on p_1, and the demand for q_2 only on p_2. The commodities are hence independent, and cannot represent substitutes or complements. This is a particularity of the Cobb-Douglas shape.

Aggregating

A particular advantage of Cobb-Douglas functions is worth noting. We can have any set of consumers, individuals identified by suffix i, with different preferences in terms of the α_1^i, α_2^i, and different incomes y^i, and yet construct a utility function for the whole group, such that the demand function can be derived as if there were only one single consumer.[12]

[10]Cobb and Douglas (1928).

[11]This is due to the assumed symmetry as we took $\alpha_1 = \alpha_2 = 0.5$. Other exponents result in other income shares, for instance $\alpha_1 = 1, \alpha_2 = 1.5$ in the shares $\frac{1}{1+1.5} = 0.4$, $\frac{1.5}{1+1.5} = 0.6$, so $q_1 = 0.4\frac{y}{p_1}$, $q_1 = 0.6\frac{y}{p_1}$.

[12]Later we will see how messy such group aggregation becomes if the demand functions were linear.

Each consumer maximizes utility $U^i = (q_1^i)^{\alpha_1^i} (q_2^i)^{\alpha_2^i}$ subject to the budget constraint $p_1 q_1^i + p_2 q_2^i = y^i$. The maximizations results in demand functions

$$q_1^i = \frac{\alpha_1^i}{\alpha_1^i + \alpha_2^i} \frac{y^i}{p_1} \tag{1.11}$$

$$q_2^i = \frac{\alpha_2^i}{\alpha_1^i + \alpha_2^i} \frac{y^i}{p_2}$$

Suppose we have n consumers. Define average income $y = \frac{1}{n} \sum_{i=1}^{i=n} y^i$ and the weighted Cobb-Douglas coefficients $\alpha_1 = \frac{1}{n} \sum_{i=1}^{i=n} \frac{\alpha_1^i}{\alpha_1^i + \alpha_2^i} \frac{y^i}{y}$ and $\alpha_2 = \frac{1}{n} \sum_{i=1}^{i=n} \frac{\alpha_2^i}{\alpha_1^i + \alpha_2^i} \frac{y^i}{y}$. Then average consumption for the whole group is

$$q_1 = \frac{1}{n} \sum_{i=1}^{i=n} q_1^i = \alpha_1 \frac{y}{p_1} \tag{1.12}$$

$$q_2 = \frac{1}{n} \sum_{i=1}^{i=n} q_2^i = \alpha_2 \frac{y}{p_2}$$

quite as if we dealt with just one single consumer whose utility was $U = q_1^{\alpha_1} q_2^{\alpha_2}$ and budget constraint was $p_1 q_1 + p_2 q_2 = y$.

1.2.3 Cobb-Douglas Coefficients and Advertising

Notably, the Cobb-Douglas exponents need not be taken as constants. Normally it is convenient to assume them as constants representing the preferences.

However, depending on the use to which we want to put the model, we might take account of the influence on tastes of actions by the firms, especially sales promotion.

To model the effect of advertising in a market with, say, two competitors supplying brands of an essentially homogenous commodity we may take

$$\alpha_1 = \frac{r_1}{r_1 + r_2} \tag{1.13}$$

$$\alpha_2 = \frac{r_2}{r_1 + r_2} \tag{1.14}$$

where r_1, r_2 denote the advertising costs of the competitors.

Such a setup catches the feature that it is the relative spending on publicity by a single firm that influences the preferences.[13]

1.3 Lancaster's New Demand Theory

If we want to model substitutes there is a tool which has been sadly neglected in "mainstream" economic theory, Kelvin Lancaster's new approach to demand theory of 1966.[14] In essence Lancaster proposed to enter not the marketed commodities themselves, but their quantified "properties" in the utility function.[15]

Thereby we get the commodities characterized through metric properties which can be a basis for classifying more or less close substitutes.

1.3.1 Properties and Commodities

Let us consider the essentials of Lancaster's model. Denote the marketed commodities q_j and their properties x_i. Further, let a_{ij} denote the quantity of property i, that the marketed commodity j contains in each unit.

Thus a desired property mix is obtained through a mix of marketed commodities:

$$x_1 = a_{11}q_1 + a_{12}q_2 \tag{1.15}$$

$$x_2 = a_{21}q_1 + a_{22}q_2$$

It is, of course, particularly nice to deal with full rank square matrices $\left[a_{ij}\right]$ because they can be inverted.

Lancaster by no means restricted the numbers of properties and commodities to be equal and as small as two, though for duopoly, to which we will apply the theory two commodities is natural.

Further, for the equality of rows and columns, there is a credible general argument: (1) If there are *more* commodities than properties, then some commodity is superfluous, i.e., dominated by a convex combination of the other ones. (2) If there

[13]This is an idea the present author once got when supervising a business economics student. It seems to be easy to set up oligopoly models, though, admittedly, it is difficult to assume nonzero production costs as well. One may argue that in oligopoly theory variable production costs were put equal to zero in order to keep the model sufficiently simple, but the defence is not entirely convincing. The idea remains largely unexplored.

[14]Lancaster (1966, 1971).

[15]Different brands of, say, cars can be characterized through their properties: Engine power, safety, comfort. If we can quantify these characteristics, the marketed goods can be represented by vectors of such properties. Just these vectors can be used to identify similar commodities, i.e., close substitutes. Once we can put value tags on the properties, we can even compare different designs.

are *less* commodities than properties, then someone can make profit from designing a new combination of properties and market it as a new product that dominates any combination of the existent.

Defining the determinant, assumed nonzero,

$$D = a_{11}a_{22} - a_{12}a_{21} \tag{1.16}$$

the equation system can be inverted:

$$\begin{aligned} q_1 &= \frac{a_{22}}{D}x_1 - \frac{a_{12}}{D}x_2 \\ q_2 &= \frac{a_{11}}{D}x_2 - \frac{a_{21}}{D}x_1 \end{aligned} \tag{1.17}$$

1.3.2 Optimum

Lancaster suggested that the utility function to be maximized should have x_i, the properties, as arguments, not q_j, the commodities, i.e.[16]

$$U = \sqrt{x_1 x_2} \tag{1.18}$$

The budget constraint, on the other hand, must contain the marketed commodities

$$p_1 q_1 + p_2 q_2 = 1 \tag{1.19}$$

Commodity prices are denoted p_1, p_2—for the properties we do not, as yet, have any prices. To avoid unnecessary symbols, income is normalized to unity.[17]

---

[16]This also opens a wider perspective—as we get shadow prices for the properties, we can also extend models to choose design for the competing products. Hopefully, in a global dynamics, we could then not only find out the orbits of changing transaction and commodity prices, but also study the design of the competing products. It would be interesting to know if the competitors would asymptotically converge upon the same design, or, rather choose very different designs. Other attractors than fixed points, for instance periodic, would be even more interesting, as one could then model fashion cycles.

[17]Economists as a rule have the habit of including as many parameters as possible, probably to attain some kind of "generality". In this book, we rather use the custom from physics to only include parameters that individually influence something in a model. For instance the budget constraint

$$p_1 q_1 + p_2 q_2 = y$$

can be written

$$\frac{p_1}{y} q_1 + \frac{p_2}{y} q_2 = 1$$

1.3.3 "Shadow" Property Prices

We now have to maximize utility $U = \sqrt{x_1 x_2}$ subject to the budget constraint $p_1 q_1 + p_2 q_2 = 1$ and the Lancaster transformations $x_i = a_{i1} q_1 + a_{i2} q_2$, $i = 1, 2$. To this end, associate Lagrange multipliers λ_1, λ_2 with the Lancaster equations. Not unexpectedly they turn out to be imputed "shadow" prices for the properties x_1, x_2.[18]

The Lagrangean optimand reads

$$\Lambda = \sqrt{x_1 x_2} + \sum \lambda_i \left(a_{i1} q_1 + a_{i2} q_2 - x_i\right) + \left(1 - p_1 q_1 + p_2 q_2\right) \tag{1.20}$$

so, maximizing with respect to x_1, x_2, we get

$$\begin{aligned} \frac{1}{2} \frac{\sqrt{x_1 x_2}}{x_1} &= \lambda_1 \\ \frac{1}{2} \frac{\sqrt{x_1 x_2}}{x_2} &= \lambda_2 \end{aligned} \tag{1.21}$$

and with respect to q_1, q_2

$$\begin{aligned} a_{11}\lambda_1 + a_{21}\lambda_2 &= p_1 \\ a_{12}\lambda_1 + a_{22}\lambda_2 &= p_2 \end{aligned} \tag{1.22}$$

The λ_i can be interpreted as values or prices of the properties. The first equation contains coefficients a_{11}, a_{21} which tell how many units of properties 1 and 2 go into each unit of commodity 1. If λ_1, λ_2 are values of the properties, then $a_{11}\lambda_1 + a_{21}\lambda_2$ tells us how much a unit of commodity 1 is worth, and this value is equated to the price p_1 of this commodity. Similarly for the second commodity.

Further, if we multiply the equations by q_1, q_2 respectively, and add, we get

$$\begin{aligned} & q_1 \left(a_{11}\lambda_1 + a_{21}\lambda_2\right) + q_2 \left(a_{12}\lambda_1 + a_{22}\lambda_2\right) \\ &= \lambda_1 \left(a_{11} q_1 + a_{12} q_2\right) + \lambda_2 \left(a_{21} q_1 + a_{22} q_2\right) \end{aligned} \tag{1.23}$$

But, using Lancaster's transformation $a_{11} q_1 + a_{12} q_2 = x_1$ and $a_{21} q_1 + a_{22} q_2 = x_2$. Hence

$$p_1 q_1 + p_2 q_2 = \lambda_1 x_1 + \lambda_2 x_2 = 1 \tag{1.24}$$

so income is just a scaling factor for prices.

[18] The budget constraint also needs a Lagrange multiplier, but it has no significance for the model and can be absorbed as a factor in λ_1, λ_2.

i.e., the original budget constraint can in optimum be written in terms of properties and their values.

1.3.4 Demand Functions

From $\frac{1}{2}\frac{\sqrt{x_1 x_2}}{x_1} = \lambda_1$ and $\frac{1}{2}\frac{\sqrt{x_1 x_2}}{x_2} = \lambda_2$ we have $\lambda_1 x_1 + \lambda_2 x_2 = \sqrt{x_1 x_2}$, so, as $\lambda_1 x_1 + \lambda_2 x_2 = 1$, we conclude that in optimum $U = \sqrt{x_1 x_2} = 1$. This in itself is not an interesting conclusion as we do not take utility as anything measurable, but used in the optimum conditions for x_1, x_2 these simplify to $\frac{1}{2}\frac{1}{x_1} = \lambda_1$ and $\frac{1}{2}\frac{1}{x_2} = \lambda_2$ or

$$x_1 = \frac{1}{2}\frac{1}{\lambda_1} \tag{1.25}$$
$$x_2 = \frac{1}{2}\frac{1}{\lambda_2}$$

which are demand functions for properties as dependent on property values. As always with Cobb-Douglas utility income is split in fixed fractions, $\frac{1}{2}, \frac{1}{2}$ in the symmetric case, so the demand for properties only depends on their own prices.

However, this is not the case for marketed goods, which is the whole point in using the Lancaster model. Substituting the property demand functions in the inverted Lancaster system we get

$$q_1 = \frac{a_{22}}{D}\frac{1}{2}\frac{1}{\lambda_1} - \frac{a_{12}}{D}\frac{1}{2}\frac{1}{\lambda_2} \tag{1.26}$$
$$q_2 = \frac{a_{11}}{D}\frac{1}{2}\frac{1}{\lambda_2} - \frac{a_{21}}{D}\frac{1}{2}\frac{1}{\lambda_1}$$

To take full advantage of the model, note that the optimum conditions for prices: $a_{11}\lambda_1 + a_{21}\lambda_2 = p_1$ and $a_{12}\lambda_1 + a_{22}\lambda_2 = p_2$, can themselves be inverted and yield

$$\lambda_1 = \frac{a_{22}}{D}p_1 - \frac{a_{21}}{D}p_2 \tag{1.27}$$
$$\lambda_2 = \frac{a_{11}}{D}p_2 - \frac{a_{12}}{D}p_1$$

So, finally substituting commodity prices p_j for imputed property prices λ_i, cancelling the determinant D in numerator and denominator, and rearranging

$$q_1 = \frac{1}{2}\frac{1}{p_1 - \frac{a_{21}}{a_{22}}p_2} + \frac{1}{2}\frac{1}{p_1 - \frac{a_{11}}{a_{12}}p_2} \tag{1.28}$$
$$q_2 = \frac{1}{2}\frac{1}{p_2 - \frac{a_{12}}{a_{11}}p_1} + \frac{1}{2}\frac{1}{p_2 - \frac{a_{22}}{a_{21}}p_1}$$

As a final step define the ratios[19]

$$b_1 = \frac{a_{11}}{a_{12}} \qquad (1.29)$$
$$b_2 = \frac{a_{22}}{a_{21}}$$

They represent, for each property, the ratios to which the two products of marketed goods contain these in their design. Substituting in the demand functions we get[20]

$$q_1 = \frac{1}{2}\frac{b_2}{b_2 p_1 - p_2} + \frac{1}{2}\frac{1}{p_1 - b_1 p_2} \qquad (1.30)$$
$$q_2 = \frac{1}{2}\frac{1}{p_2 - b_2 p_1} + \frac{1}{2}\frac{b_1}{b_1 p_2 - p_1}$$

Note that if $q_1 < 0$ then we must set $q_1 = 0$ and $q_2 = \frac{1}{p_2}$. Likewise, if $q_2 < 0$ then $q_2 = 0$ and $q_1 = \frac{1}{p_1}$. In either case only one commodity will be demanded, i.e., the stage is set for monopoly.

1.3.5 *Product Design*

Before leaving the Lancaster model, note that once we have property prices λ_1, λ_2, we can also evaluate different *new* designs in terms of the design coefficients a_{ij}

$$a_{11}\lambda_1 + a_{21}\lambda_2 \qquad (1.31)$$
$$a_{12}\lambda_1 + a_{22}\lambda_2$$

[19] They represent, for each property, the ratios to which the two products of marketed goods contain these in their design. Note that b_1, b_2 depend on design decisions by both competitors, and also that they are defined in an asymmetric way.

[20] Of course, only restricted price ratios provide for positive demand of both commodities. Most succinctly the condition can be expressed in only two compound variables, $b_1 \frac{p_2}{p_1}$ and $b_2 \frac{p_1}{p_2}$,

$$-2 < \frac{\left(b_1 \frac{p_2}{p_1} - 1\right) + \left(\frac{1}{b_2}\frac{p_2}{p_1} - 1\right)}{(b_1 \frac{p_2}{p_1} - 1)\cdot(\frac{1}{b_2}\frac{p_2}{p_1} - 1)} < 0$$

In this interval the commodities are substitutes. If any of the demand equations is negative, then only one commodity is demanded, so $q_1 = \frac{1}{p_1}, q_2 = 0$, or $q_1 = 0, q_2 = \frac{1}{p_2}$, the awkward case commented above.

These may during some evolving process become higher than the respective current commodity prices p_1, p_2. If this happens it may be appropriate to change design of the commodities. However, there are also costs associated with changing design, so we will have to defer a discussion of this topic until we have discussed the theory of production.

1.3.6 Inverse Demand and Revenue

As indicated, profit maximum in monopolistic competition is better formulated in the style of Amoroso, so we want to use the inverse demand functions

$$p_1 = \frac{1}{2}\frac{b_1}{b_1 q_1 + q_2} + \frac{1}{2}\frac{1}{q_1 + b_2 q_2} \tag{1.32}$$
$$p_2 = \frac{1}{2}\frac{1}{b_1 q_1 + q_2} + \frac{1}{2}\frac{b_2}{q_1 + b_2 q_2}$$

Note that if $b_1 = b_2 = 1$, i.e., the commodities are identical, then embeds the case of Cournot duopoly, provided. To see this, substitute to obtain

$$p_1 = p_2 = \frac{1}{q_1 + q_2} \tag{1.33}$$

References

Benthnam J (1789) An introduction to the principles of morals and legislation. Prometheus Books, Amherst
Cobb CW, Douglas PH (1928) A theory of production. Am Econ Rev 18:139–165
Jevons WS (1871) The theory of political economy. Palgrave Macmillan, Basingstoke
Lancaster KJ (1966) A new approach to consumer demand. J Polit Econ 74:132–157
Lancaster KJ (1971) Consumer demand: a new approach. Columbia University Press, New York
Menger C (1871) Grundsätze der Volkswirtschaftslehre. Braumüller, Vienna
Schumpeter JA (1954) History of economic analysis. Allen & Unwin, London

Chapter 2
Elements of Production

Abstract Like in the previous chapter we try to collect production functions that may prove useful in a global dynamic analysis. In particular we focus varieties of the constant elasticity of substitution function with negative exponents which have been hardly used at all, but which prove most useful if we want to incorporate capacity constraints in production and cost functions. Combined with "putty/clay" ideas of malleable versus fixed capital during investment periods from the 1970s this much facilitates the discussion of evolving competition among few big firms to many small firms.

Production theory contains no philosophical intricacies of kind that appear in connection with utility. The product output of a firm is measurable without the slightest doubt. Production theory in the modern sense seems to have appeared around 1820 with Johann Heinrich von Thünen and David Ricardo. Both focused on "diminishing" returns, especially in application to agriculture. Applying more and more capital and labour to a given lot of land was argued to yield less and less in additions to the crop. Thus, marginalism, decreasing returns, found its way also to the theory of production, in parallel to decreasing utility.

Later treatises show up much of formal similarities, as the most influential treatises of the twentieth century, by Sir John Richard Hicks and Paul Anthony Samuelson demonstrate.

The most important treatise particularly devoted to production theory is by first Nobel Laureate Ragnar Frisch.[1]

Later, production theory became focused in an axiomatic form (Shephard), quite like general equilibrium theory. In this form the exercises are useless for our context. Instead the useful items come from "minor" contributions in empirical production studies.

[1]Frisch (1965). Though published in French as late as 1963 and in English 1965, it was available in Norwegian as mimeographed "Memorandum fra Sosialoekonomisk Institutt" Oslo in numerous editions since 1927. Several scientists who could read Norwegian copied it and built own careers on its contents. Today we would call it plagiarism.

T. Puu, *Disequilibrium Economics*, https://doi.org/10.1007/978-3-319-74415-5_2

2.1 Inputs, Output, and the Production Function

The concept of production function tells how much product can be obtained from any given combination of quantities of factors of production when the best known technology is applied for each combination. One also speaks of inputs for the factors of production and output for product.

Such a definition guarantees free disposal of superfluous inputs. As a rule there is an idea of an optimal combination of inputs for each production process, so applying more and more of one input while keeping the other constant may result in smaller and smaller additions to the resulting output. One experiences decreasing returns. However, from a certain point on one may even reach the case where additional inputs decrease output, as when crowding of manpower at a building worksite makes efficient work more difficult. In such a case the maximum output clause assures that superfluous labour can be left unused.

However, such a definition, though logical, may be quite difficult to relate to the technical background for any real production process.

2.1.1 Decreasing Returns

Decreasing returns in production relate to two different issues. As discussed above, adding more and more of one input, for instance capital, while keeping another, for instance (farming) land, constant, adds less and less to the output (decreasing *marginal* product).

There is a general understanding that this is a fact of partial variations of one (or a few) inputs. On the other hand, there is the idea that if we, for instance, double all inputs, then output as well will be doubled. This is equivalent of setting up an additional production plant of exactly the same kind as a previous, which one thinks always must be possible. When this holds, one speaks of constant *returns to scale*. Frisch elaborates this point, though he, as a fan of fancy terms calls this the "*pari passu law*".[2]

---

[2]Frisch also has the idea that decreasing returns will not occur from the beginning when a first unit of an input is applied. First, there is an interval where marginal returns are increasing, only eventually to be reversed to decreasing returns. This is what Frisch names the "*regular ultra passum law*".

This idea lies behind all textbook discussions of production theory with its U-shaped marginal and average cost curves.

Admittedly, this is one of the most realistic ideas economics ever had. Philosophically speaking, increasing returns are the reason why things start to emerge, whereas eventually decreasing returns safeguard that they stay limited and do not grow beyond any measure. Economics, which is flooded by exploding linear models should profit from more use of this idea.

However, algebraically things in production theory become complicated, as polynomials representing production functions become cubic, with multiple profit maximum candidates that

Another use of the decreasing returns idea is when discussing the shape of the isoquants. An isoquant is a curve that represents all the combinations of inputs that all result in the same quantity of output, quite as points of the indifference curve in consumption theory result in the same utility. The difference is that output is measurable, whereas utility is not.

An isoquant is represented by a falling curve. If one wishes to keep output constant while decreasing one input, the decrease must be compensated by increasing the other. Further, successive decreases must be compensated by larger and larger increases of the other. This represents an idea that there is an optimal mix of inputs, and that deviations from that require larger and larger compensations.

The result is a curve that is convex to the origin. This is well represented by the Cobb-Douglas function

$$q = x_1^{\alpha_1} \cdot x_2^{\alpha_2} \tag{2.1}$$

where $0 < \alpha_1, \alpha_2 < 1$. If $\alpha_1 + \alpha_2 = 1$, then we have a case of constant returns. Returns to scale are decreasing when $\alpha_1 + \alpha_2 < 1$, increasing when $\alpha_1 + \alpha_2 > 1$. As we see the Cobb-Douglas function cannot represent variable returns to scale, such as in Frisch's regular ultra passum.

Though convexity of the isoquants, or decreasing *marginal rate of substitution* rests on the idea of decreasing returns, algebraically it is a matter of second order derivatives of the production function.

Some efforts have gone into measuring the convexity of isoquants through the "*elasticity of substitution*", a rather odd measure quite unlike for instance the radius of an osculating circle. The concept was introduced by Sir John Hicks in his "*Theory of Wages*" of 1932. It says how much the ratio of inputs changes along an isoquant as result of a marginal change of its slope, both changes in relative terms in order to produce an "elasticity" instead of just a derivative. It is infinite if we have $q = x_1 + x_2$ and zero if $q = \min(x_1, x_2)$, the Leontief production function.[3]

Solow in 1956 invented a clever family of production functions based on constant elasticities of substitution, called "*constant elasticity of substitution*", CES, functions. As we will see its full potential has not been used. For certain parameters it produces convex isoquants quite as in the Cobb-Douglas case.

---

have to be compared. Typically, few textbooks that present this idea in terms of graphs seem to have produced the graphs using any algebraic production function.

[3]Given the production function is $q = f(x_1, x_2)$, and denoting the first and second order derivatives f_i and f_{ij} respectively, we have

$$\sigma = -\frac{x_1 f_1 + x_2 f_2}{f_{11} f_2^2 - 2 f_{12} f_1 f_2 + f_{22} f_1^2} \frac{f_1 f_2}{x_1 x_2}$$

where σ is standard notation. Calculating σ for the Cobb-Douglas function, a good exercise, $\sigma = 1$ along the entire isoquant, and moreover also for all isoquants.

But for another parameter interval isoquants with built in capacity limits can be obtained, and for yet another interval production possibility frontiers in multiple production.

2.1.2 The CES Function

As mentioned Solow in 1956 proposed the "constant elasticity of substitution" production function

$$q^{\rho} = \beta_1 x_1^{\rho} + \beta_2 x_2^{\rho} \tag{2.2}$$

in a study of the Harrod model for economic growth.[4] He analyzes three cases of production technology, the Cobb-Douglas type, the Leontief type that we mentioned above and the obviously new CES type.

Solow is particularly interested in the case $\rho = 0.5$,

$$q = \left(\sqrt{x_1} + \sqrt{x_2}\right)^2 \tag{2.3}$$

where as in most our digressions omit the β_i coefficients which have no qualitative consequences and are there to provide for scale and asymmetry. As our units of output and inputs are not defined, they can always be rescaled to absorb such scalings and asymmetries.[5]

For the CES function there is a simple relation between the exponent and the (constant) elasticity of substitution,

$$\sigma = \frac{1}{1-\rho} \tag{2.4}$$

It is again a good exercise to derive this formula for the CES function from the formula given in a previous footnote. We note that for Solow's particular favourite with the exponent equal to one half we get $\sigma = 2$. We already noted that for the limiting Leontief case $\sigma = 0$, whereas for the Cobb-Douglas case $\sigma = 1$. Sometimes one sees the Cobb-Douglas function listed as a special case of CES with

[4] It has the property of a constant Hicksian measure of substitution elasticity over the entire function domain (the positive quadrant of x_1, x_2-space).

The contribution was popularized in a following article, Arrow et al. (1961).

It should be mentioned that the growth model proposed by Sir Roy Harrod (1948) is a Keynesian macroeconomic model based on savings and investments providing for balanced growth. So, Solow's thrust is to insert production technology in the growth process.

[5] As the use of production functions in economics is for econometric estimation, economists want to have as many coefficients as possible to provide for best fit to data. The case for theoretical use is rather the opposite.

$\sigma = 1$ and hence $\rho = 0$, but this is, of course, nonsense as $q = \left(x_1^0 + x_2^0\right)^{\frac{1}{0}}$ has no meaning.

The normal use of the CES function has been $\rho > 0$ to keep clear of the Cobb-Douglas case. It was also mentioned that another limiting case, the linear production function $q = x_1 + x_2$ has an infinite elasticity of substitution as now $\rho = 1$.

What about $\rho > 1$? For instance $\rho = 2$ gives the familiar formula $q = \sqrt{x_1^2 + x_2^2}$ for the radius of a circle. It has the wrong curvature for an isoquant, but is very useful for modelling a production possibility frontier in multi commodity production. To the knowledge of the present author, this possibility has not been used.

So, for positive ρ, we get a useful class of isoquants, when $\rho < 1$ in input space, when $\rho > 1$ in output space. To the reader knowledgeable in non-Euclidean geometry, all these cases represent nothing else than simple Minkowski metrics. The present author does not know if this nineteenth century cosmology was the true inspiration for the CES or not.

We can also consider cases with $\rho < 0$, which are no longer metrics, but provide another interesting class of functions, for instance $q = \frac{x_1 x_2}{x_1 + x_2}$ where $\rho = -1$. Arrow etc. discuss these cases and note that the isoquants are convex to the origin with asymptotes, quite as the Cobb-Douglas. But they think the axes are the asymptotes, which is erroneous. The asymptotes lie off the axes in the positive quadrant, which is interesting as it provides a possibility to model production with built-in capacity limits. For a correct presentation see Heathfield and Wibe.

Let us now take a more systematic look at the different cases.

Case $0 < \rho < 1$

This is the standard case for the CES function with isoquants convex to the origin, such as

$$q = \left(\sqrt{x_1} + \sqrt{x_2}\right)^2 \tag{2.5}$$

Solow's favourite case, or

$$q = \left(\sqrt[4]{x_1} + \sqrt[4]{x_2}\right)^4 \tag{2.6}$$

The isoquants in this function family have the normal convex look, though they always intersect the axes in the points $(0, 1)$ and $(1, 0)$, unlike the case for the Cobb-Douglas function where they go asymptotically to the axes in input space. In plain words the difference is that with the CES substitution between inputs is more efficient than in the Cobb-Douglas case, as in the first any output can be obtained with one input alone, $x_1 = 1$ or $x_2 = 1$. In the Cobb-Douglas case, reducing one input to zero always requires compensation in terms of increasing the other towards infinity.

The difference due to different ρ is that the higher ρ, the more pronounced is the convexity. In the limit, as $\rho \to \infty$ the isoquants approach the case of two line segments at right angles from the origin.

In Fig. 2.1 we illustrate a family of different *unit* isoquants.[6] In the Figure the isoquants for the family $0 < \rho < 1$ are shown in green shade.

Case $1 < \rho$

In this case the isoquants have the wrong convexity for input space, but, as mentioned, they would work fine in the space of outputs in cases of multiple production. So, also for notation we had better reverse the symbols and write

$$x = \left(q_1^\rho + q_2^\rho\right)^{\frac{1}{\rho}} \tag{2.7}$$

for instance,

$$x = \sqrt{q_1^2 + q_2^2} \tag{2.8}$$

or,

$$x = \sqrt[4]{q_1^4 + q_2^4} \tag{2.9}$$

which has a sharper bend than the radius formula.

See the red area in Fig. 2.1 for cases $1 < \rho$. The limiting case as $\rho \to \infty$ is

$$\max\left(q_1, q_2\right) = 1 \tag{2.10}$$

We can also consider multiple inputs in the left hand side, for instance

$$\left(\sqrt{x_1} + \sqrt{x_2}\right)^2 = \sqrt{q_1^2 + q_2^2} \tag{2.11}$$

which still has constant returns, as a doubling of both inputs and both outputs still fits the equation.

[6] A usual isoquant map would show level curves for $q = \left(x_1^\rho + x_2^\rho\right)^{\frac{1}{\rho}}$ at a selection of different q with ρ fixed. As we now want to compare different members of the CES family, we instead display the *unit* isoquants for each member, i.e. set $q = 1$ and let ρ vary. As always $1^\rho = 1$, we get the simpler equations

$$x_1^\rho + x_2^\rho = 1$$

for the isoquants.

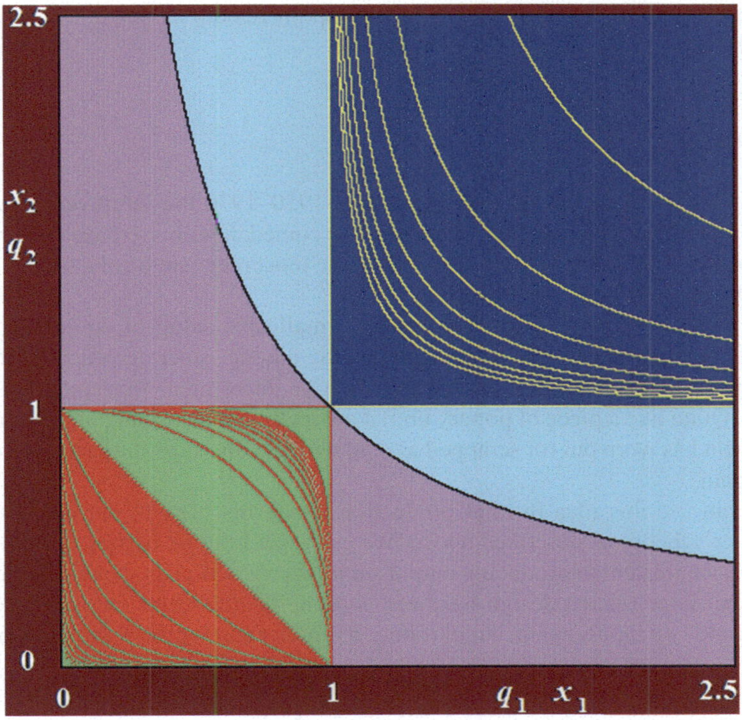

Fig. 2.1 The picture displays *unit* isoquants for a CES production function $q = \left(x_1^\rho + x_2^\rho\right)^{\frac{1}{\rho}}$. The curve families are thus not those for a given function, with fixed parameter ρ, but sets for $q = 1$ for different ρ. We can see three cases: $0 < \rho < 1$, which is the normal case with right convexity for isoquants, including Solow's favourite $\rho = \frac{1}{2}$, i.e. $q = \left(\sqrt{x_1} + \sqrt{x_2}\right)^2$. These cases are drawn as green curves on a red background. For $\rho > 1$, the convexity goes the wrong way for isoquants, but we can use it for joint production, for instance, $\left(q_1^\rho + q_2^\rho\right)^{\frac{1}{\rho}} = x$, where this single input on the right can be substituted by any input kernel, for instance, the CES Solow case, or a simple Cobb-Douglas function. The circular arc is one case for joint production $\sqrt{\left(q_1^2 + q_2^2\right)} = \sqrt{x_1 x_2}$. These cases are displayed in red against a green background. Finally, we may have $\rho < 0$, another non-utilized case. These curves are drawn in yellow on a blue background, and have asymptotes. Arrow et al. did not understand the nature of the asymptotes, they thought they were the axes, hence the same for all cases, but they are different, which makes them useful for modelling capacity limits. If one input is fixed it becomes a capacity constraint for production with the other alone. To this class belong $\rho = -1$, i.e., $q = \frac{x_1 x_2}{x_1 + x_2}$, a very simple and useful case

Such production functions can be used to optimize both the input mix to be used in production and the combination of outputs.

On Putty-Clay

In the flourishing capital and growth theory 1950–1970 the intuitively appealing "putty-clay" idea for production functions was coined. It seems to have been present in the writings of many classics already, but the most pregnant formulation is due to then 29-year old Norwegian Johansen (1959).[7]

The simple idea is that capital can be a malleable setup of equipment to be used while planning a production processes, among other technically feasible alternatives, but something unalterable once an investment in capital has been chosen. Quite like a piece of pottery until it enters the furnace. Only once the capital equipment has worn out (or scrapped as outdated), are the substitution opportunities open again.

We can use this idea in joint production which fits back to our discussion of Lancaster's theory as described above. We can regard the properties a_{ij} in product design as equivalent to producing outputs in joint production, malleable in the putty phase, but fixed once a design has been chosen. Take the illustration case of two outputs and two inputs again, but introduce a product unit for the complex of outputs

$$q = \left(\sqrt{x_1} + \sqrt{x_2}\right)^2 = \sqrt{q_1^2 + q_2^2} \tag{2.12}$$

Then set[8]

$$a_{1j} = \frac{q_{1j}}{q_j} \tag{2.13}$$

$$a_{2j} = \frac{q_{2j}}{q_j}$$

Accordingly

$$\sqrt{q_{1j}^2 + q_{2j}^2} = \sqrt{a_{1j}^2 + a_{2j}^2}\, q_j \tag{2.14}$$

[7] Johansen (1959), further elaborated in Johansen (1970). Leif Johansen was the last remarkable Norwegian economist after Frisch and Haavelmo. He died from a heart condition after a jogging round at 52 in 1982. As he took care of his autistic daughter he only spent few hours at the department. Colleagues complained, but an investigation showed that he produced about four times as much as the colleagues who were in office 9 to 5 p.m.

[8] The index j refers to the firm offering a product, but now we only consider one single firm. Note that once putty substitution is frozen at the clay level we no longer have joint production of *two* commodities q_{1j}, q_{2j} but one commodity q_j with properties a_{1j}, a_{2j}.

Recall that in the Lancaster setting we obtained imputed prices λ_i for a_{1j}, a_{2j}, so we can calculate a value for product design in terms of

$$\lambda_1 a_{1j} + \lambda_2 a_{2j} \tag{2.15}$$

and optimize with respect to a_{ij}. Of course we have a constraint. Obviously, $q_j = \sqrt{a_{1j}^2 + a_{2j}^2} q_j$, or

$$\sqrt{a_{1j}^2 + a_{2j}^2} = 1 \tag{2.16}$$

The design optimization, maximizing $\lambda_1 a_{1j} + \lambda_2 a_{2j}$ subject to $\sqrt{a_{1j}^2 + a_{2j}^2} = 1$, is quite straightforward. The optimum design, of course, concerns the composition of properties for one single unit of product (at the "putty" level). It still remains to decide optimum output q_j and input mix x_1, x_2 of the "clay" level.[9,10]

This means minimizing production cost

$$r_1 x_1 + r_1 x_1 \tag{2.17}$$

where r_j are input prices as determined on the market, subject to

$$q_j = \left(\sqrt{x_1} + \sqrt{x_2}\right)^2 \tag{2.18}$$

which is straightforward as well.

This long diversion was intended to show how product design in the Lancaster sense can be fed back from the production side. For oligopoly this allows us to compare values of design, and makes it possible to let oligopolists compete not only by prices, but also by product design. In dynamical settings we can study whether in monopolistic competition the competitors converge on identical design of their products or find their "niches" in terms of very different brands. Periodic outcomes are even more interesting as they can explain changing and recurrent designs.

[9]We refrain from product, or manufacturer, indices j when dealing with inputs x_1, x_2 in order to avoid overloading. As for the output side we considered it essential to take the full set of indices a_{ij} in order to show the connection to Lancaster's theory.

[10]Of course in order to have this optimization left we must assume that conversion from putty to clay does not eliminate substitution possibilities among inputs. However, if we prefer this as more logical, we can also fix the input mixes at this stage. As a result we get marginal costs fixed, though we still have the optimization with respect to production volume left.

Case $\rho < 0$

The third case $\rho < 0$ has been as little considered as $1 > \rho$ was. It is most interesting, though as it permits modelling with capacity limits, as represented in the yellow part of Fig. 2.1. The isoquants have asymptotes, though in the positive quadrant—not the axes as Arrow et al. thought.

To take a type case, set $\rho = 1$. Then $q^{-1} = x_1^{-1} + x_2^{-1}$ or

$$q = \frac{x_1 x_2}{x_1 + x_2} \tag{2.19}$$

so the unit isoquant is defined by

$$\frac{1}{x_1} + \frac{1}{x_2} = 1 \tag{2.20}$$

with asymptotes $x_1 = x_2 = 1$.

If any of x_1 or $x_2 < 1$, then the unit output $q = 1$ cannot be produced. Expressed in other words, if one has one input fixed at a level less than unity and yet tries to produce using only the other input, then one encounters infinite production costs. For instance if capital is fixed through an investment decision, then a capacity limit is introduced which makes it impossible to reach an output level higher than this limit using labour alone.

This is a very useful variant of the CES function as thus a smooth production function can represent built-in capacity limits. Low absolute values $|\rho|$ transfer the unit isoquant out in the positive quadrant with asymptotes at $x_1, x_2 = \frac{1}{|\rho|}$. High $|\rho|$ push it down with

$$q = \min(x_1, x_2) \tag{2.21}$$

as limiting case.

2.1.3 *Cost and Capacity Limits*

As mentioned the case $\rho < 0$ can be used to introduce capacity limits. Recall that Ragnar Frisch distinguished between cases where all relevant inputs could be varied and cases where only a single one of just a few could be varied. This latter case would always be under experience of diminishing returns, whereas the former would result in constant returns to scale, the motivation being that any completely described operation could be scaled up any number of times.[11]

---

[11]The same would hold for scaling down, though in real experience there are always indivisabilities that prevent this. Ultimately it a problem of mismatch between the world of measurements and the

Case $0 < \rho < 1$

To see the relation between partial and total variations it is more instructive to see things at the cost level. To start the discussion take the more used case $0 < \rho < 1$, for instance $q = \left(\sqrt{x_1} + \sqrt{x_2}\right)^2$. To get the cost function $C = r_1 x_1 + r_2 x_2$ when the mix of inputs x_1, x_2 is optimized given input prices r_1, r_2 we minimize C, subject to the production function $q = \left(\sqrt{x_1} + \sqrt{x_2}\right)^2$ where q is taken as a constant.

The conditions are easily obtained through the Lagrange method. The resulting minimum cost function is

$$C = \frac{r_1 r_2}{r_1 + r_2} q \tag{2.22}$$

where the quotient is a constant when input prices are given, and total cost hence proportional to output.

Likewise input demands are obtained as

$$x_1 = \left(\frac{r_2}{r_1 + r_2}\right)^2 q \tag{2.23}$$

and

$$x_2 = \left(\frac{r_1}{r_1 + r_2}\right)^2 q \tag{2.24}$$

Note that the cost function obtained is the cost function when all inputs, or production factors according to older terminology, can be varied.

It is also called the *long run* cost function, whereas the cost functions for partial variations are called *short run*, the reason being that we often consider some inputs, typically capital, as fixed in the short run through a previous investment, and variable only in a longer perspective when capital wears out. What remains for variation is labour force.[12]

So, reverting to the case of capital and labour as inputs, and using traditional symbols k for capital, l for labour, r for capital rent and w for the wage rate, we deal with the production function

$$q = \left(\sqrt{k} + \sqrt{l}\right)^2 \tag{2.25}$$

continuum, which always remains a bold abstraction. Considering upscaling, there also no doubt are physical limits for how large any operation can be made, which involves the unrealism of linear approximations to reality.

[12]However, labour force as well is usually more or less fixed by law and/or agreement with unions. Life, and not least economics, is full of simplifications. Also the classics, for instance Eugen von Böhm-Bawerk considered wage funds as part of the capital stock—not to speak of Karl Marx.

and the minimum (long run) cost function

$$C = \frac{rw}{r+w}q$$

This we contrast with the short run cost functions where labour alone is varied and capital is fixed. To derive it we first solve the production function for labour as dependent on output and capital:

$$l = \left(\sqrt{q} - \sqrt{k}\right)^2 \tag{2.26}$$

If we now substitute labour in the cost function $C = rk + wl$, then

$$C = rk + w\left(\sqrt{k} - \sqrt{q}\right)^2 \tag{2.27}$$

is the partial variation cost or short run cost function.

To find the relation between the two cost functions, take the derivative of the expression just derived and equate it to zero

$$\frac{dC}{dk} = r - w\left(\sqrt{q} - \sqrt{k}\right)\frac{1}{\sqrt{k}} = 0 \tag{2.28}$$

Solving, we have $\sqrt{\frac{q}{k}} = 1 + \frac{r}{w}$ or

$$k = \frac{w^2}{(r+w)^2}q \tag{2.29}$$

Let us now give a motivation for this procedure. First we took capital for given in the short run cost function and substituted the optimal (cost minimizing) labour force at each output given this given capital stock.

The question we just answered was: At which capital stock is any such short run function cost minimizing? If we proceed and substitute the optimal stock in the short run cost function, we get

$$C = r\frac{w^2}{(r+w)^2}q + w\left(\sqrt{q} - \frac{w}{r+w^2}\sqrt{q}\right)^2 \tag{2.30}$$

or simplified

$$C = \frac{rw}{r+w}q \tag{2.31}$$

which is the long run cost function, derived above. See Fig. 2.2 for the short and long run cost functions.

If we consider the graphics of the case, we get tangency for all the different short run functions to the long run cost (which is a straight line). There is an envelope property of the long run function—all the short run ones are tangent, but each only in one single point, otherwise the short run ones are always higher in terms of cost. The tangency points represent what Frisch calls "optimal scale".

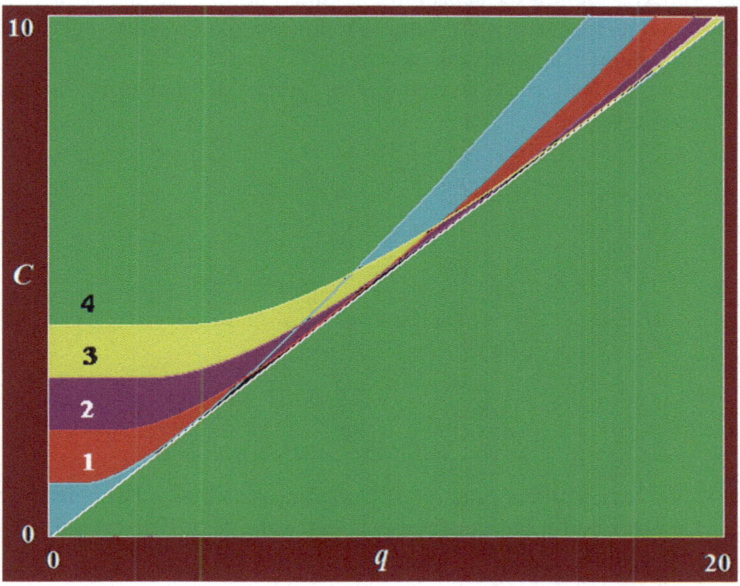

Fig. 2.2 As discussed in Sect. 2.1.1 Frisch assumes that decreasing returns are a matter of limited variability for the inputs, if all inputs are variable then the returns are constant, as any plant may simply be scaled up (or down). Though the idea is a bit unrealistic, we can illustrate it by a CES function with the parameter in the normal range. In a cost diagram marginal cost (slope of the cost function) is constant. If one input is fixed at a constant value, then marginal cost is increasing. The picture shows the constant returns cost function with a selection of four cost functions with increasing marginal cost. The relation is one of tangency, the constant returns function is an envelope to the family of decreasing returns functions. The points of tangency occur at the "optimal scale" of production, and with a finite selection of fixed inputs the output axis is divided in intervals where each member in the family is best in terms of lowest production cost

Case $\rho < 0$

Let us now revert to capacity limits. Again we take $\rho = -1$, and refer to the case of capital and labour as inputs[13]

$$q = \frac{kl}{k+l} \tag{2.32}$$

Again we minimize cost

$$C = rk + wl \tag{2.33}$$

and again there is a long and a short run alternative. Further, we still have the envelope property, but now the short run cost functions are not only always higher—they become infinite for every short run function when a capacity limit established by the size of capital is approached.

Again taking the long run first; minimize C subject to q constant. The optimization is as before, we get[14]

$$C = \left(\sqrt{r} + \sqrt{w}\right)^2 q \tag{2.34}$$

and

$$k = \frac{\sqrt{r} + \sqrt{w}}{\sqrt{r}} q \tag{2.35}$$

$$l = \frac{\sqrt{r} + \sqrt{w}}{\sqrt{w}} q \tag{2.36}$$

Notably, the long run cost function is again linear and indicates constant returns to scale. This is not so for the short run cost functions with fixed capital.

[13]That the production function now has the exact shape of the marginal cost of the previous case is purely accidental, or, at least, somehow insignificantly built into the mechanics of the function shape.

[14]Like in the previous case it turns out that the marginal cost in the long run has the same form in prices as the previous production function had in inputs. But this is no more of particular interest. Traditional economics is occupied by "dualities" between production and cost functions, but it seldom achieves more than classificatory results.

From the production function we get

$$l = \frac{qk}{k-q} \tag{2.37}$$

which substituted in the cost function gives

$$C = rk + w\frac{qk}{k-q} \tag{2.38}$$

which is the short run function. Notably, C becomes infinite whenever $q \to k$ which establishes the capacity limit. See Fig. 2.3.

We can as before equate the derivative with respect to capital to zero in order to obtain the envelope to the short term cost functions, thus

$$\frac{dC}{dk} = r - w\frac{q^2}{(k-q)^2} = 0 \tag{2.39}$$

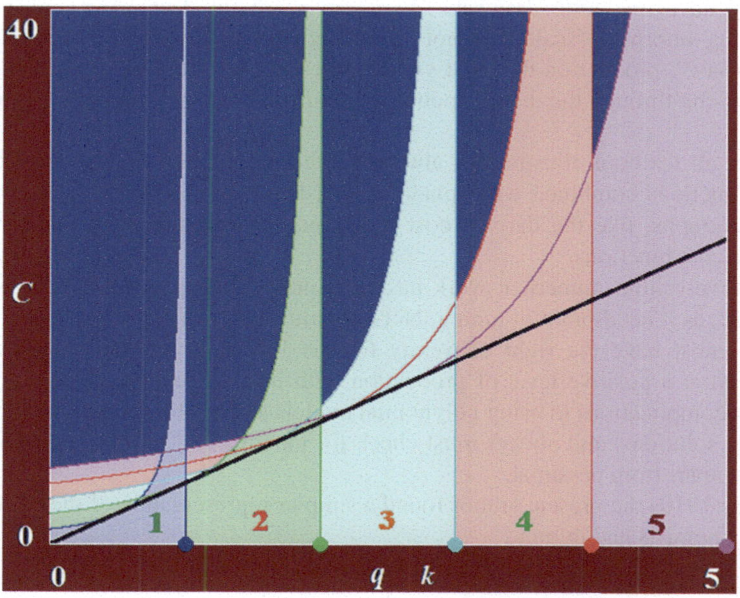

Fig. 2.3 Tangency for short run cost functions with fixed capital to the cost line when all inputs are variable still prevails. But when $\rho < 0$, then the cost functions each go to infinity when a specific limit provided by capital as capacity limit is approached

which solves for $\sqrt{\frac{r}{w}} = \frac{q}{k-q}$, or

$$k = \frac{\sqrt{r} + \sqrt{w}}{\sqrt{r}} q \qquad (2.40)$$

Substituting in the cost function we get

$$C = r\left(\frac{\sqrt{r} + \sqrt{w}}{\sqrt{r}}\right) q + w \frac{\left(\frac{\sqrt{r}+\sqrt{w}}{\sqrt{r}}\right) q^2}{\left(\frac{\sqrt{r}+\sqrt{w}}{\sqrt{r}}\right) q - q} = \left(\sqrt{r} + \sqrt{w}\right)^2 q$$

which equals the long run cost function we derived.

2.2 Returns to Scale

As we see the Cobb-Douglas and CES functions are very useful for representing production possibilities at given technological knowhow.

They have one little drawback; they can represent decreasing, constant, or increasing returns to scale, but not what Ragnar Frisch calls the "regular ultra passum law", production that first yields increasing returns, and then decreasing, after passing through the distinct point of "optimal scale" can not be modelled this way.

After all textbook illustrations always display production functions as kind of cubic functions combined with quadratic and linear terms. It is all presented in sketchy graphs, like the derived cost functions, U-shaped marginal and average variable cost functions.

Relatively little numerical work has been done using polynomials. It is a bit awkward, as it needs a lot of tuning. Not only must economic variables be positive, but we must have the right convexity for the U-shapes, get the marginal cost minimum at a positive level of production with itself a positive value etc. There are also complications in using polynomial models as one at each optimization gets multiple solutions, and always must check for maxima and minima, keep optimal solution apart from pessimal.

In the 1970s the present author found a simpler representation of the qualitative facts economists had in mind

$$q = \exp\left(-\frac{1}{x_1^{\alpha_1} x_2^{\alpha_2}}\right) \qquad (2.41)$$

where there is a simple $x_1^{\alpha_1} x_2^{\alpha_2}$ Cobb-Douglas type of kernel, but any of the CES functions would do equally well. This was never published but remained in an internal working paper, so the author thought it might be time to share the idea now. See Fig. 2.4.

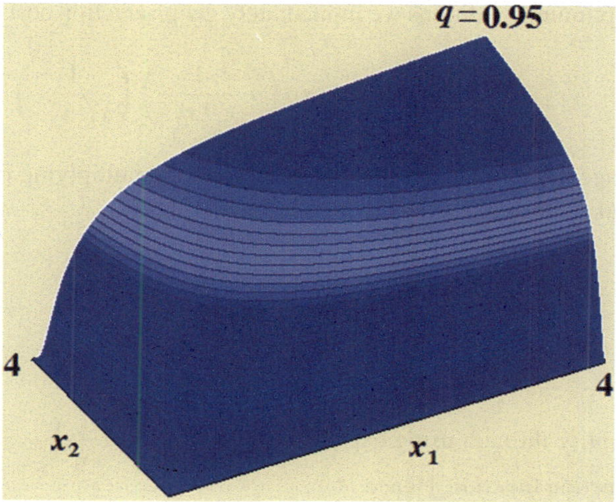

Fig. 2.4 A production function $q = \exp\left(-\frac{1}{x_1^{\alpha_1} x_2^{\alpha_2}}\right)$ with first increasing then decreasing returns to scale, corresponding to Frisch's "regular ultra passum law" and a simple Cobb-Douglas core. This is the normal textbook case though there the way of thinking is in terms of third degree polynomials which are difficult to handle due to all the constraints one has to observe

It has all the features of how economists think of production possibilities, and has the advantage of one simple formula and a certain degree of smoothness. The function has a supremum, which is attained with infinite x_1, x_2. Also, $x_1 = x_2 = 0$ result in $q = 0$. Thus the entire positive quadrant in x_1, x_2-space can be represented.

The cost functions are easy to obtain. If we minimize production cost

$$C = r_1 x_1 + r_2 x_2 \tag{2.42}$$

subject to the constraint

$$q - A \exp\left(-\frac{1}{x_1^{\alpha_1} x_2^{\alpha_2}}\right) = 0 \tag{2.43}$$

we readily get input demands

$$r_1 = \frac{\alpha_1}{x_1} \lambda A \exp\left(-\frac{1}{x_1^{\alpha_1} x_2^{\alpha_2}}\right)\left(\frac{1}{x_1^{\alpha_1} x_2^{\alpha_2}}\right) \tag{2.44}$$

$$r_2 = \frac{\alpha_2}{x_2} \lambda A \exp\left(-\frac{1}{x_1^{\alpha_1} x_2^{\alpha_2}}\right)\left(\frac{1}{x_1^{\alpha_1} x_2^{\alpha_2}}\right) \tag{2.45}$$

where λ is the Lagrange multiplier associated with the constraint.

From the optimum conditions we immediately get production cost

$$C = r_1 x_1 + r_2 x_2 = (\alpha_1 + \alpha_2)\,\lambda A \exp\left(-\frac{1}{x_1^{\alpha_1} x_2^{\alpha_2}}\right)\left(\frac{1}{x_1^{\alpha_1} x_2^{\alpha_2}}\right) \tag{2.46}$$

The Lagrange multiplier can be determined through multiplying powers α_1, α_2 of the optimum conditions and solving for the positive root

$$\lambda = \left(\frac{r_1}{\alpha_1}\right)^{\frac{\alpha_1}{\alpha_1+\alpha_2}} \left(\frac{r_2}{\alpha_2}\right)^{\frac{\alpha_2}{\alpha_1+\alpha_2}} \frac{\left(x_1^{\alpha_1} x_2^{\alpha_2}\right)^{\frac{1+\alpha_1+\alpha_2}{\alpha_1+\alpha_2}}}{\exp\left(-\frac{1}{x_1^{\alpha_1} x_2^{\alpha_2}}\right)}$$

Things simplify through using $\exp\left(-\frac{1}{x_1^{\alpha_1} x_2^{\alpha_2}}\right) = q$ and $\frac{1}{x_1^{\alpha_1} x_2^{\alpha_2}} = \ln A - \ln q$ from the production function. Hence

$$\lambda = \left(\left(\frac{r_1}{\alpha_1}\right)^{\alpha_1} \left(\frac{r_2}{\alpha_2}\right)^{\alpha_2} \left(\frac{1}{\ln A - \ln q}\right)\right)^{\frac{1}{\alpha_1+\alpha_2}} \left(-\frac{1}{q\,(\ln A - \ln q)}\right) \tag{2.47}$$

and

$$C = (\alpha_1 + \alpha_2)\left(\left(\frac{r_1}{\alpha_1}\right)^{\alpha_1} \left(\frac{r_2}{\alpha_2}\right)^{\alpha_2} \left(\frac{1}{\ln A - \ln q}\right)\right)^{\frac{1}{\alpha_1+\alpha_2}} \tag{2.48}$$

whence marginal cost

$$MC = \frac{dC}{dq} = \left(\left(\frac{r_1}{\alpha_1}\right)^{\alpha_1} \left(\frac{r_2}{\alpha_2}\right)^{\alpha_2} \left(\frac{1}{\ln A - \ln q}\right)\right)^{\frac{1}{\alpha_1+\alpha_2}} \left(\frac{1}{q\,(\ln A - \ln q)}\right) \tag{2.49}$$

and average variable cost

$$AVC = \frac{C}{q} = (\alpha_1 + \alpha_2)\left(\left(\frac{r_1}{\alpha_1}\right)^{\alpha_1} \left(\frac{r_2}{\alpha_2}\right)^{\alpha_2} \left(\frac{1}{\ln A - \ln q}\right)\right)^{\frac{1}{\alpha_1+\alpha_2}} \frac{1}{q} \tag{2.50}$$

Note that $-\frac{1}{\ln q} > 0$ as $q < 1$. As mentioned there is a supremum for $q = 1$, which in the context of costs means that the proposed production function has a built-in capacity limit, without any need for fixing any essential input such as capital. See Fig. 2.5.

MC and AVC intersect in a single point at

$$q = A \exp\left(-\frac{1}{\alpha_1 + \alpha_2}\right) \tag{2.51}$$

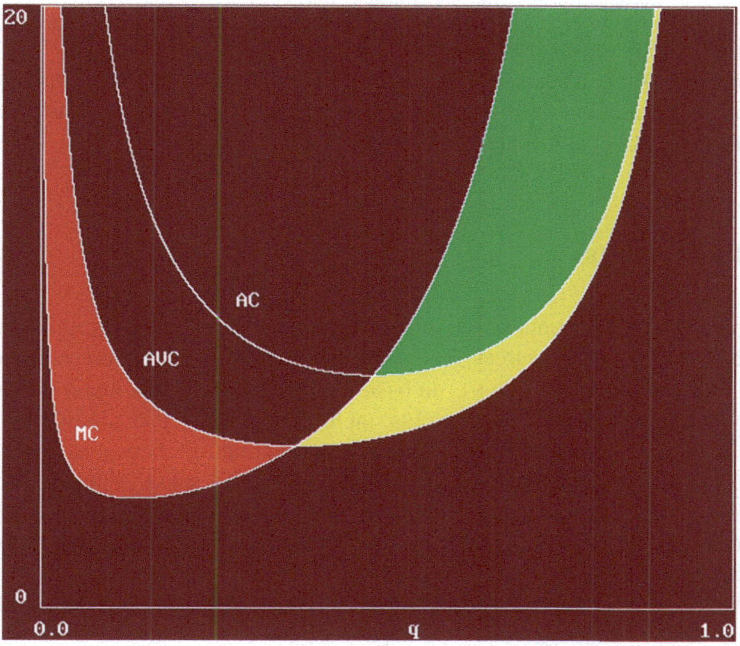

Fig. 2.5 Marginal cost $MC = \frac{2}{(\ln q)^2}$, average variable cost $AVC = -\frac{2}{q \ln q}$ and average cost $AC = \frac{1}{q} - \frac{2}{q \ln q}$ for the case $\alpha_1 = \alpha_2 = \frac{1}{2}$, $r_1 = r_2 = 1$. In the red area not even average variable costs are covered, so it is not profitable to produce. In the green area even the fixed costs are covered, whereas in the yellow proceeds suffice to cover at least part of the fixed cost, so, if they cannot be avoided it may be profitable to produce with a loss

the Frischian optimal scale, where returns are constant. It seems $q = A \exp\left(-\frac{1}{f(x_1,x_2)}\right)$ might be a useful formalization of textbook production and cost functions as it is smooth.

The alternative is third degree polynomials, with up to four parameters that all have to be constrained in intervals to produce reasonable results, and so must variables as polynomials always turn to produce wrong results outside strict intervals.

As for cost, we might find it useful to add a fixed cost C_0

$$C = C_0 + (\alpha_1 + \alpha_2)\left(\left(\frac{r_1}{\alpha_1}\right)^{\alpha_1}\left(\frac{r_2}{\alpha_2}\right)^{\alpha_2}\left(\frac{1}{\ln A - \ln q}\right)\right)^{\frac{1}{\alpha_1+\alpha_2}} \tag{2.52}$$

whence

$$AC = \frac{C_0}{q} + (\alpha_1 + \alpha_2)\left(\left(\frac{r_1}{\alpha_1}\right)^{\alpha_1}\left(\frac{r_2}{\alpha_2}\right)^{\alpha_2}\left(\frac{1}{\ln A - \ln q}\right)\right)^{\frac{1}{\alpha_1+\alpha_2}} \frac{1}{q} \tag{2.53}$$

It is likely that the fixed cost C_0 is an increasing function of the capacity limit A, as a larger plant may require more effort to get running.

A growing firm can have several plants of different capacities at its disposal, and may operate them alone or in combination, dividing production loads according to the equal marginal cost principle.

2.3 Fixed Costs

Fixed costs are of different kinds. An obvious item is any kind of capital; buildings, machines, managerial resources. These are fixed as invested resources that cannot be switched on and off on a short run basis as dependent on how much is profitable to use just then. In a sense most inputs, including contracted manpower have this character. However in theory it is tradition to distinguish between capital, land (natural resources), and labour, so we must make some sense of these categories without too much hair splitting. On the micro level we can distinguish among many different kinds of inputs, perhaps most clear in the process industries, but also visible in construction industries.

However, there is a different kind of fixed cost. The present author's supervisor called it "empty running cost", perhaps start up cost would be a better term. Most of us have experienced how just starting up a process without yet producing anything incurs considerable cost. For instance those who experienced old wax based department printers may recall the delicate start up, which in the worst case could ruin the machine, whereas once all was running everything went smoothly. It seems modern 3D printers share this property.

The author remembers a visit to a French window glass factory around 1960. There was a huge machine in a very long hall, where raw materials went in at one end and glass came out at the other. The funny thing was that the staff was occupied with drawing carriages to the exit, smashing the glass with hammers as it came out and drawing the carriages back to the entrance and putting the glass pieces back in the mouth. There was a simple explanation: The glass sheets had to be polished on six polishing machines, of which one was in repair. As it was too expensive to shut down and start up the huge machine again, smashing the output a day or two was a better choice.

However, since production theory replaced older theorizing starting from production costs and hence closer to actual technology, there was no means to incorporate such start up costs, so they just disappeared. This illustrates how theoretical paradigms shape the phenomena we perceive.

References

Arrow KJ, Chenery HB, Minhas BS, Solow RM (1961) Capital-Labor substitution and economic efficiency. Rev Econ Stat 43:225–250
Frisch R (1965) Theory of production. D. Reidel Publishing Company, Dordrecht
Harrod RF (1948). Towards a dynamic economics. MacMillan, London
Heathfield D, Wibe S (1987) Production Functions. MacMillan, Basingstoke
Hicks JR (1932) The theory of wages. MacMillan, London
Hicks JR (1939) Value and capital. Oxford University Press, Oxford
Johansen L (1959) Substitution versus fixed production coefficients in the theory of economic growth - A synthesis. Econometrica 27:157–176
Johansen L (1970) Production functions. North-Holland Publishing Company, Amsterdam
Ricardo D (1817) On the principles of political economy and taxation. John Murray, London
Samuelson PA (1947) Foundations of economic analysis, Harvard University Press, Cambridge
Shepherd RW (1952) Cost and production functions. Princeton University Press, Princeton
Solow RM (1956) A contribution to the theory of economic growth. Q J Econ 70:65–94
von Thünen JH (1826) Der isolierte Staat in Beziehung auf Landwirtschaft und Nationalökonomie. G. Fischer, Jena

Part II
Models

Chapter 3
Cournot I: Constant Returns

Abstract Cournot duopoly has now been an inspiration for mathematical modelling in economics for almost two Centuries. As the principles are very simple to grasp one needs not have a lifetime of indoctrination with economic theory to be convinced. Once nonlinear science took the stage it was mathematicians such as Rand, and Poston and Stewart, who started using applications to duopoly. Not altogether surprising, after all Cournot was a mathematician. The aim then was to create iterative processes in the style of the logistic map. Cases where both reaction functions were in this style were suggested, which, of course, could produce up to four different fixed points. Unfortunately, these were not based on economic principles, i.e., were not derived from demand functions that emerged from utility maximization. In 1991 the present author proposed the case of Cobb-Douglas utility where the resulting demand functions are always reciprocal to commodity price. Budget shares for the consumers, and hence also aggregate revenues for the producers then became constant. Dana and Montrucchio had proposed a similar case as an example 5 years before, though without fully appreciating the potential of the idea. This "isoelastic" demand function easily lends itself to explorations of dynamics. Yet, it also has its snags, above all the constancy of revenues, which leads to the absurd situation that a single supplier can reduce output to zero and sell it at an infinite price, which does not affect revenues at all. On the other hand costs can be reduced to zero by cancelling production, so the best choice is to produce nothing. This, of course, is absurd and makes the model unsuitable to deal with monopoly or collusion. The problem persists in duopoly, despite the fact that the intersection of the reaction curves in the origin is totally unstable; the curves at this intersection even have infinite slopes. Nevertheless we need some mechanism that pushes the system away from the origin if it lands there. This may sound simple, but it is not! Any such repulsion mechanism tends to take over the show, and blur the essential dynamics we want to analyze. As the reaction curves intersect the axes whereas outputs cannot be negative, it is inevitable that an occasional dropping out from production makes the system land in the origin. Once one firm chooses zero output, the other will do it as well. We need something to prevent the system from ever visiting the origin, which will be done through assuming adaptation. This means that the firms never move to the calculated best reply, just part of the way, which may be 99.99%. Given this, the assumption seems to be quite innocuous.

T. Puu, *Disequilibrium Economics*, https://doi.org/10.1007/978-3-319-74415-5_3

3.1 Introduction

3.1.1 Biographical

Antoine-Augustin Cournot (1801–1877) received his first degree in mathematics from the Sorbonne in 1823 and then further degrees in mechanics and in astronomy. He was employed for 10 years as private secretary to one of Napoleon's lesser generals, Laurent Gouvion, marquis de Saint-Cyr, created field marshal during the Russian campaign. The employer died in 1830, so this seems to be when Cournot's academic career started.

He became professor of mathematics at the University of Lyon, and later Rector of the University of Dijon. The present author did not manage to find reference to any mathematical achievement by Cournot, so it seems that his claim to fame rests on his contribution to economics of 1838, "*Recherches sur les principes mathématiques de la théorie des richesses*", addressed at the "theory of wealth". He also taught economics; one of his students was the father or Léon Walras.

3.1.2 Cournot's Contribution

There is no doubt that Cournot is the originator of oligopoly theory, and one of the founders of mathematical economics. According to Schumpeter, his agenda, addressed at markets and competition, went all way from monopoly to perfect competition, and he correctly stated that in monopoly marginal cost would equal marginal revenue, lower than price according to the formula known as Amoroso's, whereas in perfect competition marginal cost would equal price. He also showed that in duopoly with identical firms market price would be lower than monopoly price, but higher than the competitive price.

He wanted to see the whole spectrum of competitive situations due to increasing numbers of competitors. According to Schumpeter he just considered adding more and more competitors to an already existent market, which made it natural to consider a homogenous commodity and competition through supply. This made it natural to disregard the possibility of price cuts for which Cournot was later criticized by Joseph Bertrand.

Cournot always kept to what are now called naïve expectations, i.e., that the competitors observe each other's last move and react to that even if they can observe that the moves constantly change in any unfolding of a dynamic iteration. With Schumpeter we can accept this as a natural assumption, by no means self-contradictory, though not the only one possible.

3.1.3 Cournot's Heritage

Few economics models have been as successful as Cournot's, which still, almost two Centuries later continues to capture the minds of scientists. Some of its popularity is due to the fact that its basic features are so simple that researchers from other disciplines without much knowledge about economic phenomena, physicists and mathematicians, can grasp it in a moment, and use it for computations and analysis. After all Cournot himself was a mathematician. This is particularly true since complex dynamic phenomena were focused in the 1970s.

The more significant evolution of theory, however, was earlier and internal in economics. We will trace the most important steps here, even if some are discussed in entire following Chapters.

Strategic Action

A Century after Cournot Heinrich von Stackelberg in "*Probleme der unvollkommenen Konkurrenz*" 1938 made an ingenious extension of the Cournot model through suggesting that any of the competitors might try to become a "leader", through taking the reactions of the other competitor according to Cournot in explicit consideration when devising his own actions. The tenability of such a situation would, depend on whether the other was content with adhering to his Cournot-like behaviour, i.e., being a "follower".

He might take up the challenge, and try to become a leader himself. The outcome of such warfare would depend on long term production conditions for the competitors, and on their financial strengths. In the end one of the duopolists might force the other out of the market, thus becoming a monopolist.

They may agree on collusive behaviour, a possibility Cournot was aware of, provided law permits this nowadays forbidden solution. They may also return to the Cournot equilibrium, or to a Stackelberg equilibrium, provided the duopolists tacitly agree to form a leader-follower pair. As we will see it is even possible that it is the follower who profits most from the other becoming a leader, though Stackelberg was not aware of this. This may happen because the leader forces the price up through reducing supply, which both competitors profit from, at the cost of the consumers, of course.

Game Theory

At the event of game theory, duopoly was one of the obvious fields of application, and the theory was recast in terms of probabilistic strategies. This, however, falls outside the scope of the present discussion, because the points we want to make do not need anything but the classical deterministic models that arouse in the period 1838–1941.

Bertrand's Objection

A problem we consider only briefly here, as it is discussed at length below, is Joseph Bertrand's objection of 1883. According to Schumpeter his discussion was so confused that it would not have been considered by the profession if Edgeworth had not elaborated the argument a few years later. Some authors nowadays speak of Bertrand's "model"; they are advised to read the original article in "*Journal des Savants*". No such model exists, all Bertrand presented was an aggressive and confused assault. Unfortunately, the good habit of consulting original sources has fallen in oblivion, and authors are satisfied with references to reference to references, with all their accumulated modifications and misunderstandings.

In essence Bertrand's reasoning goes as follows: Cournot assumed a market for some homogenous commodity where the firms compete through supply. If so, any competitor might consider lowering its supply price. As the commodity is considered homogenous, the consumers would then, once the price difference overscores any inertia, all flow over to the cheapest supplier. This sets up an ill structured sequence of price cuts, and the only satisfactory theoretical solution is to consider the commodities supplied as different though close substitutes. The culmination of this idea was given in Edward Hasting Chamberlin's classical "*Monopolistic Competition*" of 1933.[1]

Upon the birth of studies in complex dynamics in the 1970s, the Bertrand alternative was not revived at par with the Cournot case. In the present author's opinion this has been due to the lack of convincing demand functions for substitutes.

Linear functions are always a first start, so they have been tried recently with appropriate coefficient signs, but, though linearity often is a good local approximation it never works in a global setting. Further such demand functions fail to fulfil budget constraints and can therefore not result from any normal demand theory.

Hotelling's Solution

But one case was proposed in the literature where a homogenous oligopoly could be saved from unstructured price cuts. The trick was to put the theory in a spatial setting. See Harold Hotelling (1929). Transportation costs created local monopoly areas, and suppliers at a distance could charge different mill prices for a homogenous commodity. Some competition then occurred at the market boundaries.

Unfortunately, Hotelling's main case assumed demand to be inelastic, so, as the competitors were free to choose location, they would eventually crowd in the same

[1]The subtitle "*A Re-orientation of the Theory of Value*" clearly asserts to establish monopolistic competition on equal footing with general equilibrium, as an even more general alternative theory of value.

location. The cleverly introduced space thus evaporated again, and one would be back to the original Bertrand problem.

Unfortunately, it was this case that caught the interest of the majority of Hotelling's followers, though Hotelling in a short comment correctly conjectured that the crowding would not occur if demand were elastic, as it indeed is in reality. The competitors would then, due to the contested intermediate area, locate closer than what would be socially optimal, but they would not crowd in the same point.

Though Lerner and Singer already in 1934 showed that just assuming a reservation price would blow up the case for extreme clustering, the main part of literature on Hotelling still works with the clustering case.

Stability

According to Schumpeter Cournot considered the full spectrum of markets in competition, from monopoly to perfect competition, considering addition of more and more competing firms. Nowhere does he seem to be concerned by questions of stability. The point of departure always was the linear approximation of demand functions, and marginal production costs were assumed to be constant, even zero.

Theocharis

Theocharis in a short communication 1959 considered the stability issue. In models with linear demand the slope of the marginal revenue function is always twice that of the inverse demand function, so the diagonal entries of the stability matrix are all minus one half. Accordingly, with matrices four by four, or of higher dimension, one eigenvalue then exceeds unity and thus destabilizes the system.

This became known as the "Theocharis Problem". However, Tord Palander, the present author's supervisor, demonstrated this in 1936 already at a visit to Cowles Commission. The short contribution "*Stability in competition*" was printed in its transactions. Whether Theocharis knew of this or not is something now prescribed.[2]

More interesting is that, as demonstrated by Ahmed and Agiza, the same happens to the nonlinear demand model we propose to use, only with one more competitor. So, the blame is not with demand assumptions, but with production and cost.

Constant marginal cost means constant returns to scale. A firm producing under constant returns is potentially unlimited in capacity as it can expand production with no increase in unit costs. Any positive difference between market price and marginal cost permits the firm to ripe infinite profits through expanding.

---

[2]Theocharis was an economics student at the London School of economics. In addition to this short note on stability of oligopoly he wrote a history of mathematical economics, particularly addressing the tableau économique and what later became the quantity theory of money, which became his PhD thesis.

This is no interesting situation, and definitely not what Cournot had in mind. The road from monopoly over duopoly to perfect competition should involve the comparison of few large firms to many small firms, cases that cannot be modelled with constant returns. The present author and collaborators spent some effort on this problem, and it will be dealt with extensively in Chap. 4.

Complex Dynamics

It had long been realized that the Cournot model may lead to cyclic behaviour, and David Rand in 1978 conjectured that under suitable conditions the outcome could be chaotic. The same idea occurred to mathematicians Tim Poston and Ian Stewart in their enjoyable book on catastrophe theory. These mathematical treatments, however, do not include substantial economic assumptions under which this becomes true. In what follows we supply such very simple assumptions.

Kinked Demand of Robinson and Palander

There exist known circumstances, such as piecewise marginal revenue curves with rising sections, hence producing multiple local equilibria, which also fulfil Rand's conjecture. One such case, with the demand function kinked in the Robinson (1933) sense, was dealt with by Palander in 1939.[3,4] It is easily modified to a format that

---

[3]Joan Violet Robinson (1903–1983) was a British economist at Cambridge, daughter of a general from the Boer War. Due to radical influences at Cambridge she joined the communist party and therefore did not become a full professor until she was 62. She was married to an insignificant full professor generally known as Ms. Robinson's husband. Her important contribution in this context was "*The Theory of Imperfect Competition*" of 1933, a worthy companion to Chamberlin's of the same year. In her book Robinson focuses the fact that when a new commodity is introduced it tends to be a luxury with low demand elasticity. As price is lowered it gets into the reach of the general consumer. There is a distinct threshold for this change. The models with linear inverse demand functions, restricted by the nonnegativity constraints, produce a kinked composite inverse demand and a discontinuous marginal revenue which combined with marginal cost results in multiple local profit maxima for the monopolist.

[4]Tord Folkeson Palander (1902–1972) was a Swedish professor of economics at the University of Uppsala. His first education was to an engineer, and when he later became economist he kept his habit of precision and computations. He is mainly known as a location theorist, and his thesis (in German) became a classic. He also wrote several book-length articles in Swedish, about Keynesian macroeconomics and oligopoly. In his article on oligopoly he extends the Robinson kinked demand monopoly to duopoly where the reaction functions become discontinuous, and is able to draw not only the multiple attractors or oscillating solutions, but also their basins to amazing accuracy—using just a slide ruler. As mentioned he antedated Theocharis argument by 20 years, and proposed many intriguing models in addition to the Robinson case. Though his article of 1939 is in Swedish, any student of oligopoly can be recommended to consult it as the figures contain so much information and the formulas are almost self-explanatory. The article can be downloaded from JSTORE.

increases complexity beyond that discovered by him, so as to even result in chaos. The disadvantage of these cases, however, is that solutions for the reaction functions for the duopolists become discontinuous, thus involving some extra mathematical complications.

3.2 Formal Model

3.2.1 Iso-Elastic Demand

The model we suggest is much simpler and yields explicit solutions for the reaction functions, so that we can easily focus on the dynamics of the process. It was suggested by the present author in 1991, though, a more complicated version was proposed as an example by Dana and Montrucchio in 1986.

Assume an isoelastic demand function, such that price, denoted p, is reciprocal to total demand. Demand equals supply, and the latter is made up by the supplies of the two competitors, denoted q_1 and q_2. Thus:

$$p = \frac{1}{q_1 + q_2} \tag{3.1}$$

This demand function has one problem—it does not give any reasonable solution to monopoly nor to the collusive case. The reason is that when the possibility of making total supply zero is considered, price can go to infinity and total revenue remain constant. On the other hand, total costs would then vanish, and the collusive duopolists could get the entire revenue without incurring any costs. Such a solution is purely formal and does not have any economic substance.[5]

This absurdity does not occur with any of the other cases, as the presence of a positive supply from one competitor always keeps the price finite.

However, even in oligopoly the zero output case messes things up. The formal reaction functions drop to the negative whenever the competitors supply more than a certain quantity that depends on marginal cost.

As negative supply is impossible we must replace it by zero. This means that the firm drops out and leaves the competitor as monopolist, who then reverts to the absurd solution mentioned, to sell zero output at infinite price. So, both competitors may land at zero production at some stage of the orbit. It all depends on the parameters. As we will see, in some cases the orbits will keep aloof of the origin, in other they land there, and once they do they will remain there for the rest of the orbit.

[5]Ultimately all this depends on the fact that with Cobb-Douglas utility, any consumer's good can be substituted by another while maintaining the level of utility. So, one good can go to zero and be replaced by an infinite amount of the other, despite the principle of decreasing marginal utility which holds.

Yet, the fixed point in the origin is totally unstable, as the reaction functions have infinite slope. Any slight disturbance would throw the system out from this position. However, we do not want to make the outcome depend on the exact mechanism for such disturbances, even if they always occur in reality and thus should be accounted for in any model.

This is a dilemma. The problem is that any such rejection mechanism tends to take over the show: Its rather inessential parameters, whether we use a floor at an epsilon level to replace exact zero, or use a random number generator to perturb the orbit at the origin, tend to determine the nature of the orbit, and conceal the issues we want to analyze.

Experience shows that the best strategy is to stop the system from ever going to this unstable origin, assuming adaptation, i.e., that the process does not immediately jump to the best reply, but only takes a step towards it from any previous position.

Suppose next that the duopolists produce with constant marginal costs, denoted c_1 and c_2 respectively. The profits of the two firms become:

$$V_1(q_1, q_2) = \frac{q_1}{q_1 + q_2} - c_1 q_1 \tag{3.2}$$

$$V_2(q_1, q_2) = \frac{q_2}{q_1 + q_2} - c_2 q_2 \tag{3.3}$$

The first firm maximizes $V_1(q_1, q_2)$ with respect to q_1, the second $V_2(q_1, q_2)$ with respect to q_2. Equating the partial derivatives to zero, we can solve for the reaction functions:

$$q_1 = \sqrt{\frac{q_2}{c_1}} - q_2 \tag{3.4}$$

$$q_2 = \sqrt{\frac{q_1}{c_2}} - q_1 \tag{3.5}$$

A check of the second order conditions, in fact, testifies that we always deal with local profit maxima, provided quantities are positive as indeed they should be. The reaction functions are displayed in Fig. 3.1. Their general outline is that they start at the origin, have unique maxima, and drop to zero again. The off origin intersection is unique.

Note also the green rectangle that extends from the maximum of one reaction function. If its second iterate after reflection in the other reaction function does not yet hit the reaction function in the negative, then any process of iterations of the original reaction functions (3.4)–(3.5) avoids zero, i.e., is contained in the green rectangle. Otherwise, we will need adaptation to avoid the complication mentioned.

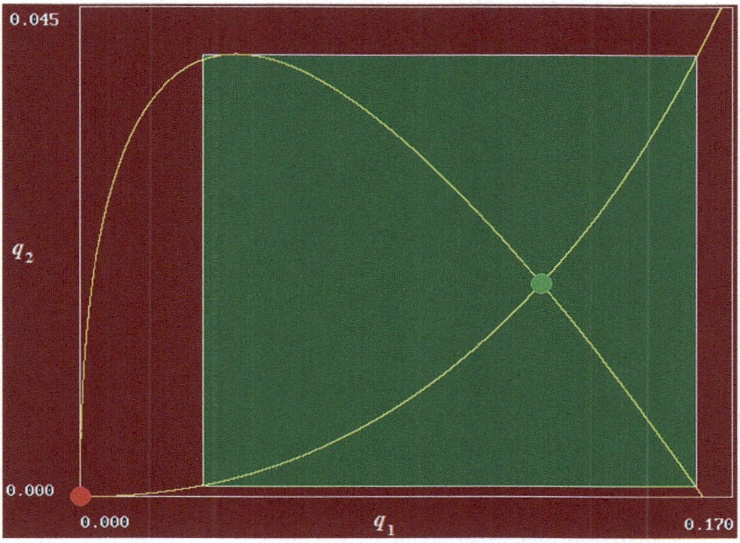

Fig. 3.1 The reaction functions for the two competitors and their intersection the green Cournot point. Obviously they cut the axes when $q_1 > \frac{1}{c_2}$ or $q_2 > \frac{1}{c1}$ but we do not show these segments as outputs cannot be negative. Some orbits, depending on the parameters, do not visit these negative segments, for instance the case displayed. The second competitor never produces more than the maximum of its unimodal reaction function, and when this maximum is reflected in the reaction function of the first firm it does not yet hit the point where the following reaction becomes negative. The orbit will be contained in the green rectangle. In this way the orbit is confined to the positive quadrant. If this does not hold special precautions must be taken to prevent the system from hitting the origin point marked in red

We can easily solve for the output quantities in the Cournot point, taking (3.4)–(3.5) as a simultaneous system of equations:

$$q_1^* = \frac{c_2}{(c_1 + c_2)^2} \tag{3.6}$$

$$q_2^* = \frac{c_1}{(c_1 + c_2)^2} \tag{3.7}$$

This point, of course, is the intersection of the reaction curves as shown by the green dot in Fig. 3.1. The profits of the duopolists at the Cournot point can be calculated by substituting back from (3.6)–(3.7) in (3.2)–(3.3):

$$V_1^* = \frac{c_2^2}{(c_1 + c_2)^2} \tag{3.8}$$

$$V_2^* = \frac{c_1^2}{(c_1 + c_2)^2} \tag{3.9}$$

3.3 The Iterative Process

Let us now turn to the dynamics. Once we are interested in the process of adjustment, we have to lag the variables. So, (3.4)–(3.5) must be written:

$$q_1' = f_1(q_2) := \sqrt{\frac{q_2}{c_1}} - q_2 \tag{3.10}$$

$$q_2' = f_2(q_1) := \sqrt{\frac{q_1}{c_2}} - q_1 \tag{3.11}$$

This pair of equations is the central piece of the iterative process. However, we have to deal with two special questions:

First, we must recognize that there is a problem with (3.10)–(3.11) as stated, when $c_1q_2 > 1$ or $c_2q_1 > 1$, because then $q_1' < 0$ or $q_2' < 0$ would be returned. Mathematically such negative outputs even result in positive maximum profits, as negative costs then dominate over negative revenues, but in terms of economics this is nonsense, and has to be ruled out. Further, any negative output would in the next iteration return a complex number due to the square root. So we rephrase (3.10)–(3.11) as:

$$q_1' = F_1(q_2) := \begin{cases} f_1(q_2) & c_1q_2 \leq 1 \\ 0 & c_1q_2 > 1 \end{cases} \tag{3.12}$$

$$q_2' = F_2(q_1) := \begin{cases} f_2(q_1) & c_2q_1 \leq 1 \\ 0 & c_2q_1 > 1 \end{cases} \tag{3.13}$$

It is a matter of the parameters c_1 and c_2 whether the entire orbit can be calculated, using the regular branches of the reaction functions (3.10)–(3.11) only, or whether the zero branches of (3.12)–(3.13) have to be included as well. To be precise, it is a matter of the marginal cost ratio c_2/c_1 alone. As we will find later, many other properties too depend on this ratio. As a matter of fact, the whole process could have been cast in terms of just one ratio parameter. But for the sake of symmetry we yet keep both.

The maximum points for $f_1(q_2)$, $f_2(q_1)$ can be easily calculated from (3.10)–(3.11) through taking the derivatives, equating to zero, and solving. Substituting back in (3.10)–(3.11), we then easily obtain the respective maximum values: $f_1\left(\frac{1}{4c_1}\right) = \frac{1}{4c_1}$ and $f_2\left(\frac{1}{4c_2}\right) = \frac{1}{4c_2}$. As we see, both maxima are located on the diagonal in phase space as shown in Fig. 3.1. Whether the process visits the zero branches or not is a matter of whether these maximum values, iterated in the other reaction function of the pair (3.10)–(3.11), lead to values that exceed the values for which (3.10)–(3.11) return a negative value or not.

As we see $f_2\left(\frac{1}{4c_1}\right) = \frac{1}{2\sqrt{c_1c_2}} - \frac{1}{4c_1}$ and $f_1\left(\frac{1}{4c_2}\right) = \frac{1}{2\sqrt{c_1c_2}} - \frac{1}{4c_2}$. In order to avoid the zero branches, these values must be less than the values for which the reaction functions (3.10)–(3.11) become zero, i.e. $\frac{1}{c_1}$ and $\frac{1}{c_2}$respectively. Accordingly:

$$\left(\frac{c_1}{c_2} \le 6.25\right) \wedge \left(\frac{c_2}{c_1} \le 6.25\right) \tag{3.14}$$

must hold in order that the process (3.10)–(3.11) be well defined even without the zero branches.

There is also a second issue which we have to settle. In earlier literature great emphasis was put on whether duopolists react simultaneously or sequentially according to a pair of reaction functions such as (3.10)–(3.11). In order to deal with this, first note that, substituting (3.11) in (3.10), or substituting (3.10) in (3.11), we obtain

$$q_1'' = f_1\left(f_2\left(q_1\right)\right) = f_1 \circ f_2\left(q_1\right) \tag{3.15}$$

$$q_2'' = f_2\left(f_1\left(q_2\right)\right) = f_2 \circ f_1\left(q_2\right) \tag{3.16}$$

These are independent iterations, each in one phase variable alone, though based on a period twice as long as the fundamental period introduced from the outset. Given we have initial phase variable values for two subsequent periods, we can compute the complete trajectory for each variable. Obviously the functions $f_1 \circ f_2$ and $f_2 \circ f_1$ are autonomous iterations, so we get the same trajectories whether the firms adjust simultaneously or in sequence. The difference between simultaneous and sequential adjustment hence only lies in the way of pairing the two iteration series together. There also is a difference in terms of the freedom to choose initial conditions. If the adjustment is sequential, then the second iterate can be calculated from only one initial condition, and we cannot choose all initial points in the phase space as depicted in Fig. 3.1, just points located on the graphs of the reaction functions.

This being said, it should be noted that the system may itself settle on sequential adjustment, even if we assume simultaneous. The union of the graphs of the reaction functions in Fig. 3.1 is an example of an invariant set. Any iteration starting on any of these leads to a trajectory which never leaves the union set. Moreover, it is attractive from a basin of nonzero area measure.

3.4 Stability of the Cournot Point

The only fixed point of this iteration, except the origin, is, of course, the Cournot equilibrium point (3.6)–(3.7). To find out something about the stability of this point we calculate the derivatives of the functions (3.10)–(3.11):

$$f_1'(q_2) = \frac{1}{2}\frac{1}{\sqrt{c_1 q_2}} - 1 \tag{3.17}$$

$$f_2'(q_1) = \frac{1}{2}\frac{1}{\sqrt{c_2 q_1}} - 1 \tag{3.18}$$

At the Cournot point with coordinates (3.6)–(3.7), expressions (3.17)–(3.18) simplify to:

$$f_1'(q_2^*) = \frac{c_2 - c_1}{2c_1} \tag{3.19}$$

$$f_2'(q_1^*) = \frac{c_1 - c_2}{2c_2} \tag{3.20}$$

Loss of stability for the Cournot point occurs when:

$$\left|f_1'(q_2^*) \cdot f_2'(q_1^*)\right| = \left|\frac{c_2 - c_1}{2c_1} \cdot \frac{c_1 - c_2}{2c_2}\right| = 1 \tag{3.21}$$

Expanding (3.21), we get:

$$(c_1 - c_2)^2 = 4c_1c_2 \tag{3.22}$$

Observe that this condition can be solved for the ratio:

$$\left(\frac{c_1}{c_2} = 3 \pm 2\sqrt{2}\right) \vee \left(\frac{c_2}{c_1} = 3 \pm 2\sqrt{2}\right) \tag{3.23}$$

Thus, whenever any ratio of the marginal costs of the duopolists falls outside the interval bounded by the two values specified in (3.23), the Cournot point will not be stable. The two roots happen to be reciprocal, so there is nothing odd in stating this condition for any of the ratios. Other critical conditions as well depend on the ratio of marginal costs only, so there is just one single free control parameter.

3.5 Periodic Points and Chaos

As this control parameter passes either of the critical values specified in Eq. (3.23), the fixed point is replaced by a two period cycle. This should be so understood that q_1 and q_2 each oscillate between two values. The two variables themselves, of course, take on different pairs of values in the oscillatory process.

After the first cycle appears, there is a period doubling cascade to chaos.

In Fig. 3.2 we illustrate the bifurcation diagram. As we see from (3.23), interesting things begin to happen when the ratio c_2/c_1 exceeds the critical value $3 + 2\sqrt{2} \approx 5.8584$, or falls short of its reciprocal. Until then there is just a stable fixed point. On the other hand, the smooth functions (3.10)–(3.11) alone are relevant if and only if c_2/c_1 does not exceed the value 6.25, or fall short of its reciprocal, as we see from (3.14). Outside this range, the zero branches of the reaction functions start being visited, and new phenomena come into play. To avoid this we limited the bifurcation diagram to the range of marginal cost ratios between 5.75 and 6.25.

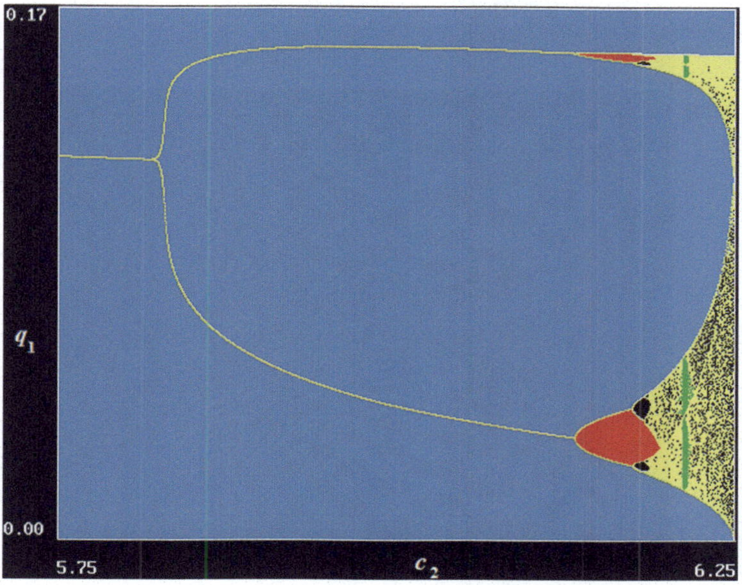

Fig. 3.2 The bifurcation cascade for the first competitor as dependent on the marginal cost for the second, i.e., the cost ratio as we keep its own marginal cost fixed at unity. The other competitor has a similar bifurcation cascade. The way to produce such pictures is to simply run the iteration from some initial condition, always trashing iterates before convergence is attained. After that the same phase value is hit over and over. Doing this for nearby parameter points produces a curve if the phase points converge to fixed points. As we see the main frame has a chaotic appearance, though interrupted by windows of order. We coloured two such big widows in red and green respectively to show where phase diagrams displayed in the following diagrams were taken

In this range we see the fixed point undergoing a period doubling bifurcation cascade to chaos. We depict the asymptotic behaviour of the q_1 variable against the marginal cost ratio c_2/c_1; the q_2 variable undergoes a similar bifurcation cascade. As already noted, the behaviour of the system is symmetric as to the ratio c_2/c_1 and its reciprocal c_1/c_2.

So shifting to the latter ratio would give the same bifurcation diagram. For the same reason, we need not extend the parameter ratio axis to the origin, because its leftmost part would only be a compressed mirror image of the information already shown.

The general appearance of the bifurcation diagram is very like that of the extensively studied logistic map. There is the period doubling cascade to chaos, as well as the windows of order in the chaotic region.

That much for the independent iterations (3.15)–(3.16). If we study the phenomena for both variables in the phase space shown in Fig. 3.1, then they combine to produce coexistent cycles. Figures 3.3 and 3.4 show two such cases. In Fig. 3.3, we took a parameter ratio $c_2/c_1 = 6.15$, which, as we can see produces 4-period cycles in each variable alone. They obviously combine to produce two different coexisting 8-period cycles. The locations of the four independent periodic points

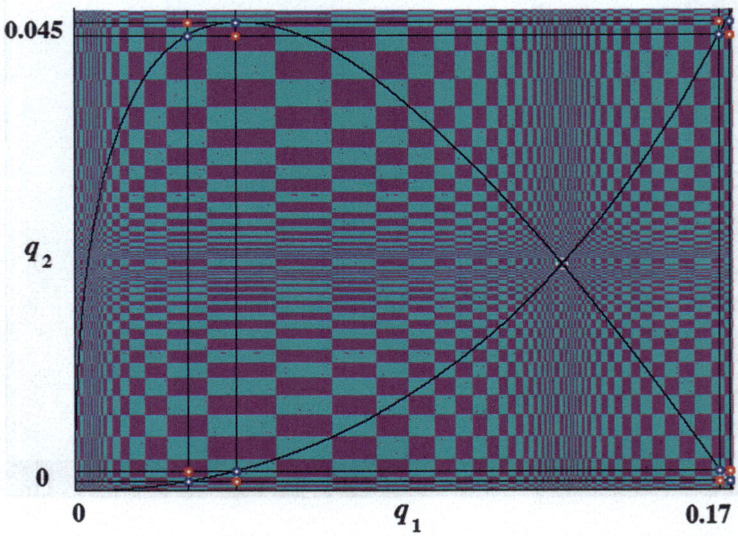

Fig. 3.3 Two coexistent 8-period cycles and their basins of attraction, one sequential with points located on the reaction functions (SEQ), the other with points off these (SIM). The parameter ratio is $\frac{c_2}{c_1} = 6.15$. In earlier studies these were assumed to depend on how the model was set up. Now we know that the system autonomously settles on one of these, which depends on the initial condition. As we see the basin is fragmented in a quite fine meshed chessboard pattern, particularly in the neighbourhood of the Cournot equilibrium point. Noteworthy is also that the attractor points of both cycles are located in the intersections of 4 horizontal and 4 vertical lines, i.e. 16 points, which also equals multiplicity of attractors, 2, times the period, 8

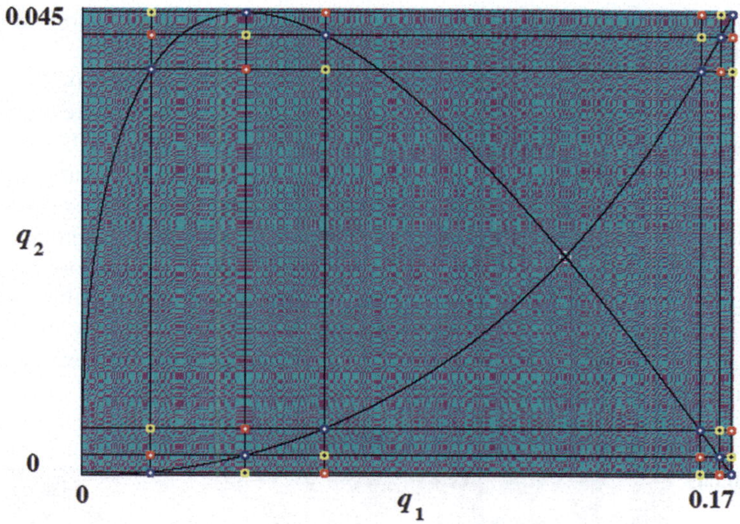

Fig. 3.4 Three coexistent 12-period cycles and their basins of attraction, now with parameter ratio $\frac{c_2}{c_1} = 6.212$. The basin mesh is so fine that things are difficult to see. Yet the same algebraic equality holds; there are 6 horizontal and 6 vertical lines producing 36 intersection points, also equal in number to the multiplicity of attractors, 3, times the period, 12. Now there is only one SEQ cycle with points on the reaction curves, whereas there are two SIM cycles with their points off these curves. As for the fragmented basin, we actually should have three colours, but we separated only the SEQ cycle basin, and merged the two SIM cycle basins

in each variable alone are indicated by four horizontal and vertical lines, whose intersections produce a 4 by 4 lattice. As we see the periodic points are located on this 16-point lattice. We also see there is a deformed checkerboard pattern of fragmented attraction basins corresponding to these coexistent cycles.

The case in Fig. 3.4 is taken for $c_2/c_1 = 6.2125$, which comes from the largest window of order we see in Fig. 3.2. In this window (3.15)–(3.16) produce 6-period cycles for each variable alone. These are again indicated by the six horizontal and vertical lines. Their intersection forms a 6 by 6 lattice. Located on it are three different sets of dots indicated by different characters, in all three coexisting 12-period cycles. The system of coexistent basins is now even more fragmented, though we only separated the sequential cycle (called SEQ from now on) from the two simultaneous (called SIM), so there are two colours as two basins are merged, instead of three. The reader can easily cook a computer program that separates all three, though the picture would become even more cluttered.

We see that in each case, if there are n-period cycles for each variable alone (n even), then they combine to produce $n/2$ different coexisting $2n$-period cycles. That this always is the case with maps such as (3.10)–(3.11) has been rigorously proved.

See Bischi, Mammana, and Gardini, where the corresponding results for odd-period n can also be found.

The bifurcation sequence leading to chaos, obviously has been of the flip, or period-doubling type. In the sequel we will encounter different bifurcations when the Cournot point loses stability. Another typical feature of the (3.10)–(3.11) type of map is the fragmented attraction basins that entirely consist of rectangular areas, getting smaller and smaller in the neighbourhood of the (unstable) Cournot point, so producing a very complex fragmentation, particularly close to this point.

As we see in both Figs. 3.3 and 3.4, the periodic points cluster in four rectangles close to the corners of phase space. When the process is chaotic, these become chaotic rectangles dense of attractor points. We see such a case in Fig. 3.5. Eventually, as the parameter ratio increases towards 6.25, these grow, and finally merge into one single chaotic rectangle.

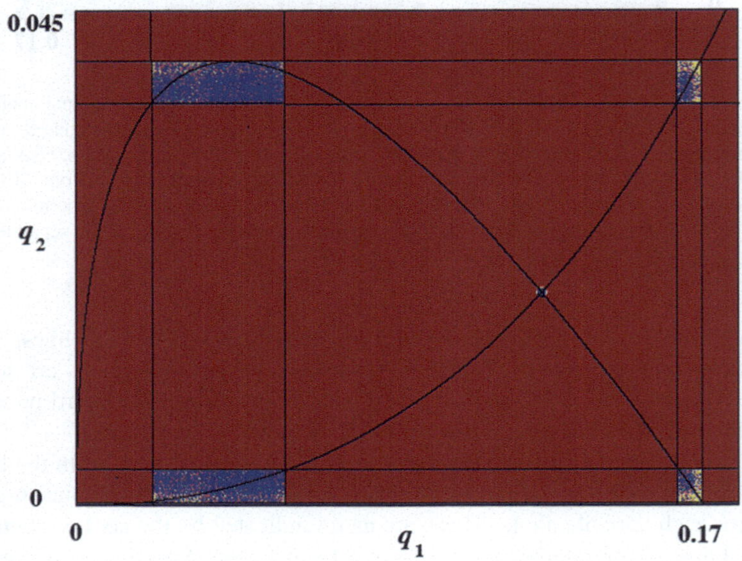

Fig. 3.5 Four piece chaotic attractor when the parameter ratio is $\frac{c_2}{c_1} = 6.2$. Note that this ratio is between those represented in the two previous pictures. It seems that all pictures came from an essentially chaotic regime, the previous are exceptional cuts from narrow periodic windows in parameter space. Note that it is the rectangles that now are the attractor, whereas the basin is the entire plane displayed. The colouring in the rectangles represent counts of the frequency of visits from the orbit, yellow visited very often, blue but seldom. Experiment shows that increasing the parameter ratio essentially enlarges the four rectangles until they merge to cover the entire picture. This route is, however, interrupted by infinitely many windows of order with predominantly high period cyclic attractors and extreme fragmentation of the basin. There is a rectangular pattern in all these pictures, which, however changes radically once we introduce adaptation

3.6 Adaptive Expectations

As mentioned, the iterations considered up to now are actually two independent ones. To make the map really two-dimensional, and take precautions to avoid the origin, we next assume that the duopolists do not immediately jump to their new optimal positions, but adjust their previous decisions in the direction of the new optimum. Assume:

$$q_1' = \phi_1(q_1, q_2) := (1-\delta_1)\, q_1 + \delta_1 \left(\sqrt{\frac{q_2}{c_1}} - q_2\right) \tag{3.24}$$

$$q_2' = \phi_2(q_1, q_2) := (1-\delta_2)\, q_1 + \delta_2 \left(\sqrt{\frac{q_1}{c_2}} - q_1\right) \tag{3.25}$$

With the adjustment speeds unitary, we are back at the case already treated. In the contrary case, with those speeds zero, the duopolists never revise any decision taken. Intermediate values, between zero and unity, bring in a host of new possibilities.

Again we must deal with the problems arising when the expressions within large parentheses become negative. So, we redefine the system as:

$$q_1' = \Phi_1(q_1, q_2) := \begin{cases} \phi_1(q_1, q_2) & c_1 q_2 \le 1 \\ (1-\delta_1)\, q_1 & c_1 q_2 > 1 \end{cases} \tag{3.26}$$

$$q_2' = \Phi_2(q_1, q_2) := \begin{cases} \phi_2(q_1, q_2) & c_2 q_1 \le 1 \\ (1-\delta_1)\, q_2 & c_2 q_1 > 1 \end{cases} \tag{3.27}$$

It is easily checked that the adaptive system (3.24)–(3.25), or (3.26)–(3.27), has the same fixed points as the original system without adaptation (3.10)–(3.11), i.e. the, always unstable, origin, and the Cournot equilibrium point, whose coordinates were given in (3.6)–(3.7). The Cournot point may again be stable, or it may lose stability. The stability of various solutions, such as the Cournot fixed point, and the cycles, now depend, not only on the marginal cost ratio c_2/c_1, but on the adjustment speeds δ_1, δ_2 as well.

3.7 The Neimark Bifurcation

Compared to the previous model, the stability loss will now be through a Neimark-Sacker bifurcation. We can study the stability around any point in phase space by linearizing the system (3.24)–(3.25). The matrix of the linearized system then becomes:

$$J(q_1, q_2) = \begin{bmatrix} 1-\delta_1 & \delta_1 \left(\frac{1}{2\sqrt{c_1 q_2}} - 1\right) \\ \delta_2 \left(\frac{1}{2\sqrt{c_2 q_1}} - 1\right) & 1-\delta_2 \end{bmatrix} \tag{3.28}$$

Considering the Cournot point, we substitute for the phase variables from (3.6)–(3.7). The matrix then becomes:

$$J\left(q_1^*, q_2^*\right) = \begin{bmatrix} 1-\delta_1 & \delta_1 \frac{c_2-c_1}{2c_1} \\ \delta_2 \frac{c_1-c_2}{2c_2} & 1-\delta_2 \end{bmatrix} \tag{3.29}$$

At the threshold of loss of stability for the Cournot point the determinant of the matrix (3.29) becomes unitary. Putting $\left|J\left(q_1^*, q_2^*\right)\right| = 1$, and simplifying, we get the condition:

$$(c_1 - c_2)^2 = 4c_1c_2\left(\frac{1}{\delta_1} + \frac{1}{\delta_2} - 1\right) \tag{3.30}$$

Note that, with the adjustment speeds unitary, condition (3.30) becomes identical with (3.22), as it indeed should.

The characteristic equation is now easily obtained from the Jacobian matrix (3.29) by putting:

$$\begin{vmatrix} 1-\delta_1-\lambda & \delta_1 \frac{c_2-c_1}{2c_1} \\ \delta_2 \frac{c_1-c_2}{2c_2} & 1-\delta-\lambda \end{vmatrix} = 0 \tag{3.31}$$

where λ are the eigenvalues. Written out (3.31) becomes:

$$\lambda^2 - (2-\delta_1-\delta_2)\lambda + (1-\delta_1)(1-\delta_2) + \delta_1\delta_2\frac{(c_1-c_2)^2}{4c_1c_2} = 0 \tag{3.32}$$

We easily recognize the coefficient of the linear term as the negative of the trace, and the constant term as the determinant of the Jacobian matrix (3.29).

Solving (3.32) we obtain the eigenvalues:

$$\lambda_{1,2} = \frac{1}{2}(2-\delta_1-\delta_2) \pm \frac{1}{2}\sqrt{\delta_1^2 + \delta_2^2 - \delta_1\delta_2\frac{c_1^2+c_2^2}{c_1c_2}} \tag{3.33}$$

To see whether they are real or complex on the bifurcation curve, we substitute from (3.30) for the quotient under the root sign in (3.33), obtaining:

$$\lambda_{1,2} = \frac{1}{2}(2-\delta_1-\delta_2) \pm \frac{1}{2}\sqrt{(\delta_1+\delta_2)(\delta_1+\delta_2-4)} \tag{3.34}$$

As the remaining parameters are adjustment speeds and only can take values in the unit interval, i.e. $0 \leq \delta_1, \delta_2 \leq 1$, we conclude that the expression under the root sign must be negative at the moment of bifurcation. The eigenvalues are hence complex conjugates, and the bifurcation accordingly of the Neimark-Sacker type.

At the moment of bifurcation the complex eigenvalues cross the unit circle in the complex plane, so we can also write (3.34) as:

$$\lambda_{1,2} = \cos\theta \pm i \sin\theta \tag{3.35}$$

where, as usual, $i = \sqrt{-1}$, and

$$\cos\theta = \frac{1}{2}\left(2 - \delta_1 - \delta_2\right) \tag{3.36}$$

$$\sin\theta = \frac{1}{2}\sqrt{\left(\delta_1 + \delta_2\right)\left(4 - \delta_1 - \delta_2\right)} \tag{3.37}$$

The bifurcation can be to a periodic or to a quasiperiodic/chaotic orbit, depending on whether θ is a rational or an irrational multiple of 2π.

From numerical work we will find that the Neimark bifurcation for the present model is of the subcritical type, not of the supercritical. The subcritical bifurcations are more difficult to understand than the supercritical, and the theorem about bifurcation of the fixed point to a closed invariant curve, whenever strong resonance is absent, does no longer apply. As a consequence, the fixed point may bifurcate to a periodic orbit, not only at the sparse rational points on the bifurcation curve, but over entire intervals.

As a matter of fact the Arnol'd tongues protrude through the bifurcation curve, and are thick already at the moment of the Neimark bifurcation. In the supercritical bifurcation, the attractive invariant curve is born the moment the fixed point is destabilized. In the subcritical, the fixed point is destabilized by an unstable invariant curve (or some other unstable set) collapsing around the point and so destabilizing it. This means that the unstable set was created through another bifurcation, together with a stable one, even before the Neimark bifurcation. After the unstable set destabilizes the fixed point, its stable companion attracts the phase points which formerly went to the fixed point. See Guckenheimer and Holmes or Kuznetsov for more detail on subcritical Neimark bifurcations.

Meanwhile, between the two bifurcations, we can find coexistence of fixed point and some other attractor, something which does not occur with a supercritical bifurcation.

Though the periodic Arnol'd tongues do not grow out from the Neimark bifurcation curve, we can still locate the periodic tongues by substituting various rational multiples of 2π for θ in the eigenvalues. So, put:

$$\theta = 2\pi \frac{m}{n} \tag{3.38}$$

in (3.36), and solve for:

$$\delta_1 + \delta_2 = 2\left(1 - \cos\left(2\pi \frac{m}{n}\right)\right) \tag{3.39}$$

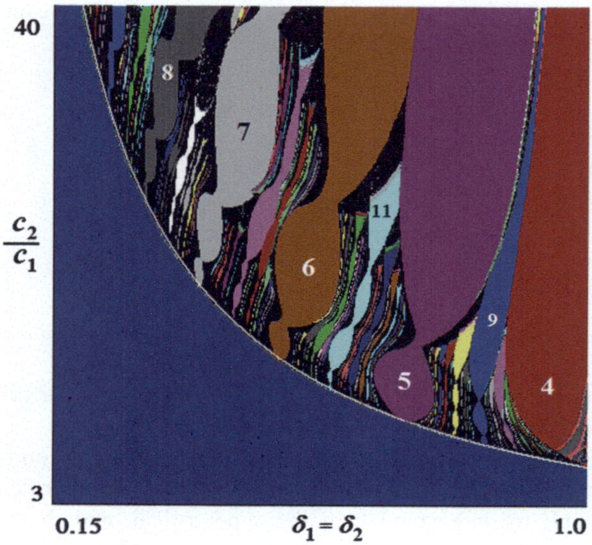

Fig. 3.6 Neimark bifurcation curve and periodic Arnol'd tongues. We actually have three free parameters, the price ratio and two adaptation coefficients, but displaying in three dimensions is tricky, so we set the adaptation coefficients equal. The way the picture is then produced is then explained in the BASIC program appended in this chapter. The periodicity points on the bifurcation curve are calculated explicitly, as is the bifurcation curve itself. Note that the numerically obtained tongues penetrate slightly through the bifurcation curve, which is an indication of subcriticality of the bifurcations, meaning coexistence of attractors, and bifurcation through disappearing basins. Noteworthy is also the strange twisted character of the Arnol'd tongues, probably due to applications of different regimes of the map

Taking first $m = 1$, we get: for $n = 1$, $\delta_1 + \delta_2 = 0$, for $n = 2$, $\delta_1 + \delta_2 = 4$, for $n = 3$, $\delta_1 + \delta_2 = 3$, and for $n = 4$, $\delta_1 + \delta_2 = 2$. We find that there is no strong resonance, except the only possibility $\delta_1 = \delta_2 = 1$ for the non-adaptive case. Though, we cannot invoke the theorem that the bifurcation necessarily is to a closed invariant curve, this gives an indication that the lowest resonance we may find is $n = 4$. The next higher harmonics are $\delta_1 + \delta_2 = \left(5 - \sqrt{5}\right)/2 \approx 1.382$ for $n = 5$, and $\delta_1 + \delta_2 = 1$ for $n = 6$, which are perfectly in the admissible parameter range.

In Fig. 3.6, we display the bifurcation diagram, which confirms that the corresponding resonance tongues indeed are located as expected from the above calculations. However, note that we have three parameters, the marginal cost ratio, c_2/c_1, and the adjustment speeds δ_1 and δ_2. In order to make it possible to draw a plane diagram, we must reduce their number to two. Numerical experiment indicates that the best choice is to put $\delta_1 = \delta_2$, and to use this along with the marginal cost ratio c_2/c_1.

In the picture we can identify, from right to left, the tongues of periodicities $n = 4$ and higher, quite as we can infer from the above calculations. The biggest tongues are indicated through numbers. There is one series running from 4 to 8, the following ones are too thin for numbering. However, we also note larger tongues numbered, 9, 11, 13, and 15, interspersed between the tongues we already spoke of.

The reason for this is that in the above calculations we took $m = 1$, so the resonances represented were 1 : 4, 1 : 5, 1 : 6, etc. But these are the same as 2 : 8, 2 : 10, 2 : 12, etc., so it is not surprising that we can find resonances 2 : 9, 2 : 11, etc. in between. Actually, with equal adjustment speeds, Eq. (3.39) reads

$$\delta_1 = \delta_2 = 1 - \cos\left(2\pi \frac{m}{n}\right) \tag{3.40}$$

so for $m = 1, n = 4$ we have $\delta_1 = \delta_2 = 1$, and for $m = 1, n = 5, \delta_1 = \delta_2 \approx 0.691$. Calculating $\delta_1 = \delta_2 = 1 - \cos(4\pi/9) \approx 0.826$ for $m = 2, n = 9$, we find that indeed the 2 : 9 tongue fits between the 1 : 4 and 1 : 5 tongues.

In all, we checked for the 45 lowest periods in Fig. 3.6. As the tongues get thinner and thinner, we conjecture that in this manner we have obtained most of the information about periodicities. It should be said that we used the system (3.24)–(3.25), including the zero branches in calculating the periodicities. The "bowel like" structure in the upper parts of the picture is probably due to the application of the zero branches. Note also that the tongues penetrate through the bifurcation curve, as expected with subcriticality.

As a companion to Fig. 3.6, we also present Fig. 3.7, which displays the areas of positive Lyapunov exponents in the parameter plane in bright shade. Note how this

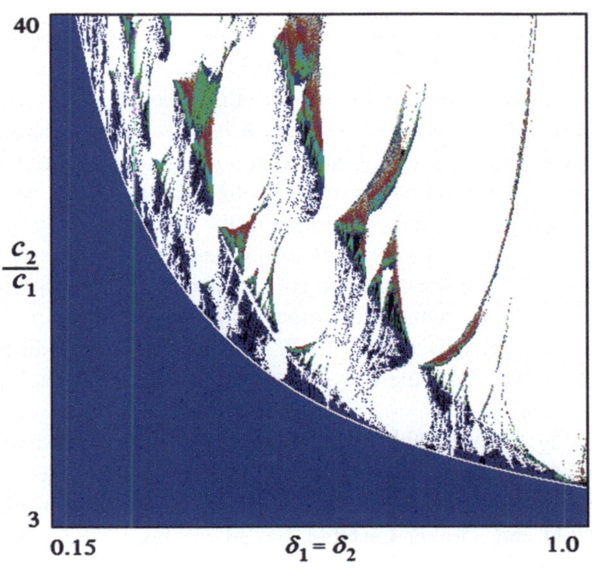

Fig. 3.7 Areas of positive (coloured) and negative (white) Lyapunov exponents. Note how this picture would fit into the blanks left by the Arnol'd's tongues in the previous, indicating how chaotic regimes alternate with periodic. Overlaps are possible though when chaos coexists with periodicity, of which we will soon see an example. The calculation of the largest Lyapunov exponent is explained in the BASIC program appended to this Chapter

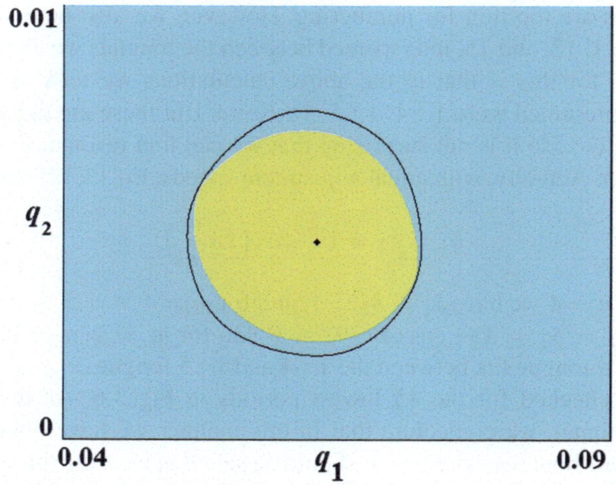

Fig. 3.8 Coexistent fixed point and invariant "curve" with parameters $\delta_1 = \delta_2 = 0.5$, and $\frac{c_2}{c_1} = 13.7635$. The parameter point is located just below the bifurcation curve close to the periodic point (in this case period 6). There is hence coexistence of the invariant chaotic curve with the Cournot fixed point. Note that such an invariant curve is not a continuous curve, such as appears in differential equation systems at a Hopf bifurcation. It only has the appearance of a curve in the resolution of the screen. It is generated through showers of scattered points. The yellow basin of the central fixed point extends almost to the invariant curve, which indicates that it was newly created. Changing parameters towards the bifurcation point eventually annihilates the stability of the fixed point through shrinking the yellow basin around it

picture of the possibly chaotic region fits into the areas where no periodic tongues are present in Fig. 3.6. We use two different bright shades, depending on whether the Lyapunov exponent is only positive, or larger than 0.01, in order to distinguish quasiperiodic and high period cycles from chaotic motion (white). Also note that the area for positive Lyapunov exponents, quite as the low periodic tongues in Fig. 3.6 did, protrudes right through the Neimark bifurcation curve.

In Figs. 3.8 and 3.9 we illustrate the coexistence of the fixed point and another attractor just before the Neimark bifurcation. As we know, the 6-period tongue starts out at the parameter value $\delta_1 = \delta_2 = 0.5$. To get the corresponding value for the marginal cost ratio we substitute this in (3.31), whence we can solve for $c_2/c_1 = 4\sqrt{3} + 7 \approx 13.9282$.

As we see in Fig. 3.6, the stable fixed point for parameter combinations just under the bifurcation curve, seems to coexist with another attractor. So, for Fig. 3.8 we choose a slightly lower $c_2/c_1 = 13.7635$, which is just after a pair of invariant curves, one stable and another unstable, have been born. The stable curve starts

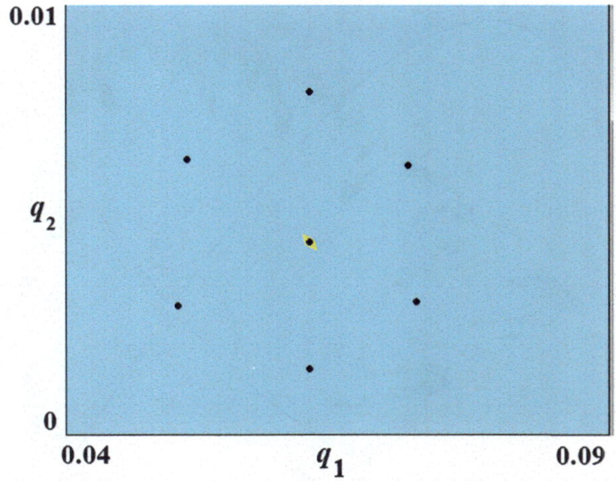

Fig. 3.9 This is the sequel of the story from the previous picture. The yellow basin of the fixed point has shrunk to almost nothing, so it will in the next move annihilate its stability. We also see that the chaotic invariant curve meanwhile has broken up in six disjoint points, i.e., a 6-period cycle. This means that the parameter point moved from a chaotic region to an Arnol'd tongue, but the scenario described happens below the hyperbola shaped bifurcation curve, in the region of periodic and chaotic tongues penetrating through the bifurcation curve. This is but a simple illustration to what can happen before a subcritical bifurcation. For those interested in a fuller scenario the studies by Professor Anna Agliari cannot be too highly recommended

getting a hexagonal feature as we are in the 6-period range, the unstable one is the boundary of the yellow basin of attraction for the fixed point.

Figure 3.9 is drawn much closer to the Neimark bifurcation curve. We put $c_2/c_1 = 13.9245$. As we see the basin of attraction for the fixed point has shrunk together to a hardly visible lemon around the fixed point, so as to annihilate its stability in the next moment. Moreover, we see that the closed invariant curve has undergone a global saddle-node bifurcation to a 6-period cycle, as indicated by Fig. 3.6, where the 6-period tongue penetrates through the bifurcation curve.

We finish this discussion by displaying a chaotic attractor in the shape of a "leaf" in Fig. 3.10, where there is no longer any coexistence with a fixed point. The parameter values are $\delta_1 = \delta_2 = 0.9$, $c_2/c_1 = 6.9262$. A cascade of dark green points show divergence from the neighbourhood of the unstable Cournot equilibrium, which ultimately under 100,000 iterations aggregate to outline the edges of the "leaf".

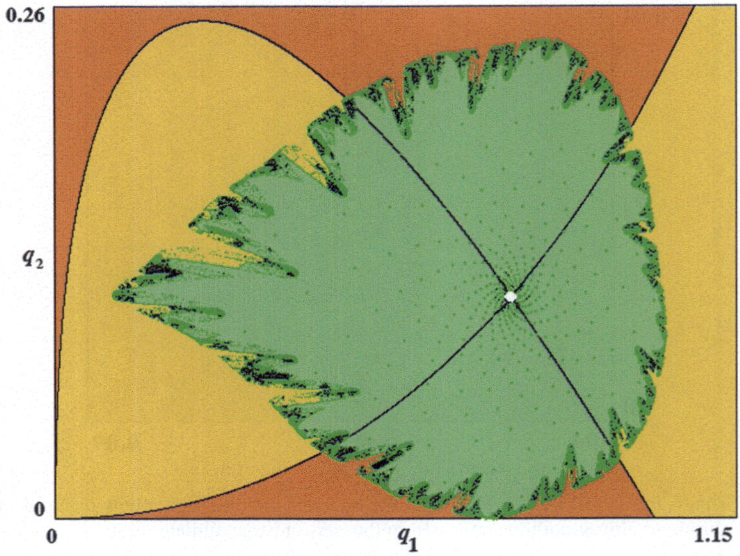

Fig. 3.10 Fractal attractor of the adaptive model. Parameters: $\delta_1 = \delta_2 = 0.9$, $\frac{c_2}{c_1} = 6.9262$. As indicated before the attractors and basins change radically with the introduction of adaptation. After a fixed point bifurcates to an invariant curve, it starts to deform, and can show up interesting shapes such as this "leaf". Note that the attractor itself just outlines its edges, the green inside colour is just for fun to emphasize the shape. In darker shade we see the initial divergence from the neighbourhood of the unstable Cournot point

3.8 Critical Lines and Absorbing Area

We can now investigate the case with the method of critical lines and absorbing areas. See Abraham, Gardini, and Mira or Mira, Gardini, Barugola, and Cathala about this remarkable method. A very simple introduction can also be found in Puu (2000, 2003). To outline the shapes of attractors we need the critical line L_{-1}.

We obtain the equation for L_{-1}, putting the determinant of the Jacobian matrix (3.28) equal to zero. Observe that we now take the general matrix (3.28), not the one at the Cournot point (3.29). Also note that we put the determinant equal to zero, not to unity as in deriving the Neimark bifurcation curve (3.30), the reason being that now we just want to locate the curve along which the map folds the plane.

So, putting the determinant of (3.28) equal to zero, and deleting the time period index for the phase variables, we obtain the following expression:

$$\left(\frac{1}{2\sqrt{c_1 q_2}} - 1\right)\left(\frac{1}{2\sqrt{c_2 q_1}} - 1\right) = \frac{(1-\delta_1)(1-\delta_2)}{\delta_1 \delta_2} \tag{3.41}$$

It is easy to see that this implicit equation represents a pair of hyperbolas in the q_1, q_2-plane. Their exact forward images according to (3.26)–(3.27) are too complex

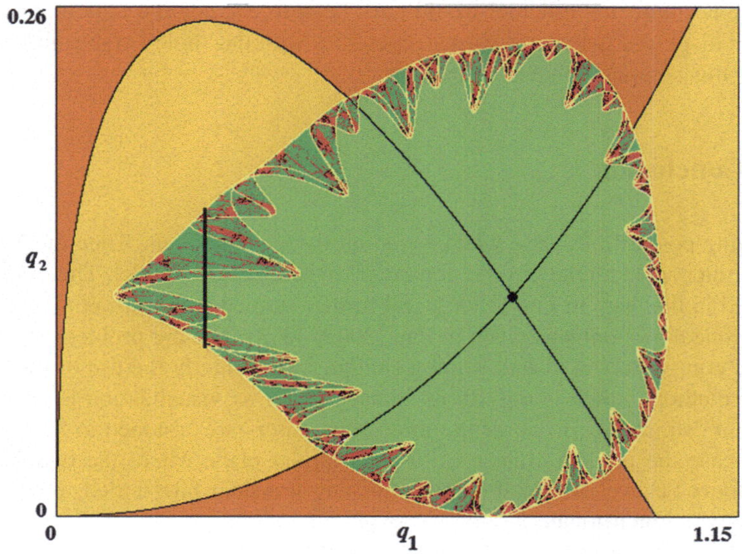

Fig. 3.11 Obtaining the absorbing area through iterating the vertical line segment on the left—actually a segment of an almost vertical piece of a hyperbola—50 times. The absorbing area closes up with a much smaller number of iterates (yellow critical lines), though the outline of the inner hole takes more iterates. As before $\delta_1 = \delta_2 = 0.9$, $c_2/c_1 = 6.9262$. This remarkable method of critical lines makes it possible to outline very complex attractors by tightly wrapped absorbing areas, and to study their bifurcations systematically. The idea is illustrated in the BASIC computer programme appended. However, it may be tricky to use as one needs a piece of a curve from the degenerate set of points of the map intersecting the attractor, as short as possible. This makes one wonder whether it is simpler to use than iterating the map to find its asymptotic set of points. Despite all its elegance, probably not. The method has a curious likeness to artists Christo and Jeanne-Claude wrapping of real objects, such as the Pont Neuf or the Berlin Parliament

to write down, but they can be easily traced by the computer. Further, the numerical case illustrated in Fig. 3.11 has $\delta_1 = \delta_2 = 0.9$. Hence the right hand side of is (3.41) is small, of the order of magnitude 0.01. If it were zero then either of the brackets on the left would have to be zero. The pair of hyperbolas then becomes a cross of straight lines. This is almost true with small but nonzero right hand sides, at least with a certain distance from the intersection of this "cross" (the hyperbola asymptotes).

In Fig. 3.11, the relevant part of L_{-1}, i.e., that which intersects the attractor, is a vertical line segment as far as we can see in these pictures. It corresponds to the second bracket in (3.41) being zero, i.e., to $q_1 = \frac{1}{4c_2}$. As we remember, this coordinate value represents the maximum of the reaction function for the second firm.

In Fig. 3.11 we also show 50 forward iterates of this segment, which obviously suffice to outline the attractor. We can easily recognize the attractor from the previous picture which has been superposed and note how well it is wrapped up by the train of critical lines. We also show the reaction functions and the unstable Cournot point at their intersection.

More accurate and efficient would have been to exclude the middle part of the segment displayed, which, as we see, coincides with the "hole" in the attractor, but we skip this computational complication.

3.9 Conclusion

Interesting possibilities arise when we admit three oligopolists. Then the Neimark bifurcation occurs already in the non-adaptive setup of the model. The model was extended in this way in Puu (1996), and further studied in Agliari et al. (2000). A less technical account was given in Puu (2000, 2003). The big problem in the case of three competitors is visualization techniques, particularly the use of the critical surface method, which is only in its infancy as far as visualization is concerned. Things are simplified if two of the three competitors are assumed to be identical. In that case the process converges to an invariant plane where the two identical competitors behave identically, and, after any transient has settled, phase space remains two dimensional.

One of the most interesting issues in connection with oligopoly models of this type are the border collision bifurcations which precede the subcritical Neimark bifurcation, as this is mathematically fairly unexplored ground. The present author had the privilege of collaborating with more knowledgeable friends, Professors Anna Agliari and Laura Gardini in some pioneering work in 2004–2005 listed in the references, in which professor Agliari contributed some as intriguing as beautiful artwork.

3.10 Computer Programs

Understanding nonlinear dynamical systems is largely a matter of experimental work at the computer. In the following we give some hints for how to write computer programs that produce some of the pictures that illustrate this chapter. The code is given in Quick Basic Version 4.5. This is rather old fashioned, as it emerged with the simplified FORTRAN version called BASIC many years ago. It reappeared a couple of years ago with a new efficient C compiler, and is free to download from the web. The disadvantage is that the graphics is still VGA, 640 by 480, with 16 colours. The advantage is that one can rely on the simple logic of `FOR` `TO` `STEP` and `IF` `THEN` `ELSE` and similar instructions, to which all programming can eventually be reduced. One also has no need for learning all the complicated input and output rules for lists of various kinds.

In what follows we give information for three different program types; drawing bifurcation diagrams with periodic Arnol'd tongues in parameter planes, drawing areas for positive largest Lyapunov exponent in the same type of planes, and drawing pictures in the phase planes using the method of iterated critical lines LC for finding shapes of attractors.

3.10.1 Drawing Periodic Tongues

In this program we calculate the lowest periodicity for each point of the parameter plane. It is good to exemplify it here, even if it is used over and over in the book, because the iteration itself (3.26)–(3.27) is so simple.

Note that overlaps of tongues cannot be seen as we choose colour according to the lowest period detected. The first block again sets the resolution and window, and draws a rectangle for the displayed picture. First, however the maximum dimensions of the vectors of stored iterates for periodicity checking are set.

Next block prints a list of the colour code for identification of the periodicities. Next block calculates and draws the Neimark bifurcation curve.

The final large block checks the periodicities. From a given initial point X_i, Y_i, 916 iterates are calculated. Note that when calculating the iterates, we use the zero branches of the reaction functions setting the values equal to a small positive number rather than to zero. The reason is that zero is interpreted as an exact number by the computer, so that the computer may stick at the unstable fixed point at the origin. The small additions actually represent random disturbances present in reality, but not in the world of the computer. After trashing the first 901 transient iterates, the last 15 are stored. A procedure then checks the distance of the last iterate from the next but last, and so on, chooses a colour according to the lowest periodicity number for which the iterates are very close, and paints a new pixel in the parameter plane. Then the procedure goes to the next parameter combination, and so forth.

First we need to store variables q_1, q_2 while checking for periodic return to some point. As we only have 16 colours including black, 15, or 30 (in the second round), etc. are chosen, but to have a safe margin we set the vector dimension to at most 100:

```
DIM q1Chk(100), q2Chk(100)
```

Next we need to set the points to check in the parameter plane. We only have two dimensions in the plane, so we set adaptation coefficients equal, $\delta_2 = \delta_1$. As for marginal costs we know that only their ratio counts for the dynamic, so we fix $c_1 = 1$, and choose the other c_1 for display. Hence we define the loops, taking generous ranges for the parameters and small steps. With a modern computer we do not need to economize:

```
c1=1
FOR c2 = 4 TO 40 STEP 0.005
FOR del1 = 0.15 TO 1 STEP 0.0015
del2 = del1
```

Once all intermediate computations are done, we have to end the loops by statements:

```
200 NEXT del1
NEXT c2
```

For running the iteration we need an initial point in the phase space, so set

```
q1i = 0.00001: q2i =0.00001
```

For the iteration, experiment shows that less than 1000 iterations provide for convergence, so, for checking the 15 lowest periodicities, we set the number of iterations at 916, trashing the first 900, so

```
q1 = q1i: q2 = q2i
FOR i = 1 TO 916
q1Nw = del1 * q1 + (1-del1) * (SQR(q2/c1)-q2)
IF q2*c1 > 1 THEN q1Nw = (1-del1) * q1
q2Nw = del2 * q2 + (1-del2) * (SQR(q1/c1)-q1)
IF q2*c1 > 1 THEN q2Nw = (1- del2) * q2
q1 = Xnew: q2 = q2new
```

This goes on during the trashed first 900 iterations:

```
IF i < 901 THEN GOTO 100
```

After that the iterates are stored, with redefined numbers 1 to 16

```
j = i - 900
q1chk(j) = q1: q2chk(j) = q2
100 NEXT i
```

The last task is to check for lowest periodicity and choosing a colour point for the parameter plane

```
FOR i = 2 TO 16
colour = 0
chk = ABS(q1chk(i)-(q1chk(1)))
+ ABS(q2chk(i)-(q2chk(1)))
IF check < 0.0001 THEN colour h-1
IF colour > 0 THEN PSET (del1,c2), colour
GOTO 200
NEXT i
```

This completes the program, of course the graphics output screen, the window size and the like have to be set. Note that we used `colour` as a variable for colour; `COLOR` (in American spelling) is a command and cannot be used in variable manipulations.

3.10.2 Positive Lyapunov Exponent

The first block is quite the same as for the periodicity diagram:

```
c1=1
FOR c2 = 4 TO 40 STEP 0.005
FOR delta1 = 0.15 TO 1 STEP 0.0015
del2 = del1
```

followed by

```
200 NEXT del1
NEXT c2
```

The same holds for the iteration

```
FOR i = 1 TO 1000
q1Nw = del1*q1+(1-del1)*(SQR(q2/c1)-q2)
IF q2*c1 > 1 THEN q1Nw =(1-delta1)*q1
q2Nw = del2*q2+(1-del2)*(SQR(q1/c1)-q1)
IF q2*c1 > 1 THEN q2Nw =(1- del2)*q2
q1 = Xnew: q2 = q2new
```

This time we do not need any storage vectors, as each iterate just modifies the estimate for the largest Lyapunov exponent. We need to keep track of two items, the direction of elongation `VCOS,VSIN` in the phase plane, and the sum of logarithmic expansions `LYSUM`. For the start we set

```
VCOS = 1/SQR(2):VSIN = 1/SQR(2)
LYSUM = 0
```

In each step the matrix of map the Jacobian is calculated

```
J11 = 1 - del1
J12 = del1 * (1/(2*SQR(c1*q2)-1)
J21 = del2 * (1/(2*SQR(c2*q1)-1)
J22 = 1 - del2
```

and the derivatives, projections on the direction cosines,

```
Dx = J11 * VCOS + J12 * VSIN
IF c1 * q2 > 1 THEN Dx = J11 * VCOS
Dy = J21 * VCOS + J22 * VSIN
IF c2 * q1 > 1 THEN Dy = J22 * VSIN
```

Then the largest Lyapunov estimate is updated

```
DSQR = Dx * Dx + Dy * Dy
AMPL = SQR(DSQR)
LYSUM = LYSUM + LOG(AMPL)
LYAP = LYSUM / i
```

And the direction cosines

```
VCOS = Dx / AMPL
VSIN = Dy / AMPL
```

Once the iteration comes to the end for each parameter point we check `LYAP` for positivity and set a point in the parameter space

```
IF LYAP > 0 THEN PSET (delta1, c2)
```

3.10.3 Drawing Critical Lines and Absorbing Areas

This time we again find it convenient to use vectors for the forward iterates of point of the critical lines, so we set the maximum number generously at 150.

```
DIM q1(150), q2(150)
```

Further set the parameters

```
c1 = 0.14488: c2 = 0.98945
del1 = 0.9: del2 = 0.9
m = 1000:
```

For reference then draw the reactions functions (main branches)

```
FOR q2 = 0 TO 1/c1 STEP 0.0001
q1 = SQR(q2/c1) - q2: PSET (q1, q2): NEXT q2
FOR q1 = 0 TO 1/c2 STEP 0.0001
q2 = SQR(q1/c2) - q1: PSET (q1, q2): NEXT q1
```

and the attractor from a convenient initial point

```
q1 = 0.95: q2 = 0.145
FOR i = 1 TO 15000
q1nw = (1-del1)*q1 + del1*(SQR(q2/c1) - q2)
q2nw = (1-del2)*q2 + del2*(SQR(q1/c2) - q1)
IF i < m THEN GOTO 100
PSET (q1nw,q2nw), 12
100 q1 = q1nw: q2 = q2nw
NEXT i
```

We are now ready to draw the critical lines. For this we need a piece of curve intersecting the attractor on which the Jacobian is degenerate according to (3.41). This must be found by trial and error. For a too short interval stretch the train of lines does not close up, for a too long the picture becomes blurred by irrelevant curves. As we know, the curves of degeneracy are a pair of hyperbolas, but in the interval of intersection with the attractor the piece of hyperbola is almost a vertical line, so, we just need the relevant stretch:

```
u0 = 0.98945: u1 = 0.9929
```

Next calculate the coordinates of the initial critical line segment

```
FOR u = u0 TO u1 STEP 0.000001
v = ((1 - 1/SQR(u))/(K - 1/SQR(u)))^2
q1(0) = u/(4*c2): q2(0) = v/(4*c1)
```

using

```
K = (del1 + del2 - 1)/(del1 * del2)
```

and we are ready to calculate and draw all the forward images of its points:

```
FOR i = 0 TO 50
q1(i+1) = (1-del1)*q1(i) + del*(SQR(q2(i)/c1) - q2(i))
q2(i+1) = (1-del2)*q2(i) + del2*(SQR(q1(i)/c2) - q1(i))
PSET (q1(i+1),q2(i+1)), 14
NEXT i: NEXT u
```

and we are done.

References

Abraham RH, Gardini L, Mira C (1997) Chaos in discrete dynamical systems. Springer, Berlin

Agiza NH (1998) Explicit stability zones for Cournot game with 3 and 4 competitors. Chaos, Solitons Fractals 9:1955–1966

Agliari A, Gardini L, Puu, T (2000) The dynamics of a triopoly Cournot game. Chaos, Solitons Fractals 11:2531–2560

Agliari A, Gardini L, Puu T (2005a) Global bifurcations in duopoly when the Cournot point is destabilized through a Subcritical Neimark bifurcation. Int Game Theory Rev 8:1–20

Agliari A, Gardini L, Puu T (2005b) Some global bifurcations related to the appearance of closed invariant curves. Math Comput Simul 68:201–219

Ahmed E, Agiza NH (1998) Dynamics of a Cournot game with *n* competitors. Chaos, Solitons Fractals 10:1179–1184

Bertrand J (1883) Théorie mathématique de la richesse sociale. J des Savants 48:499–508

Bischi GI, Mammana C, Gardini L (2000) Multistability and cyclic attractors in duopoly games. Chaos, Solitons Fractals 11:543–564

Cournot A (1838) Récherces sur les principes mathématiques de la théorie des richesses (Paris)

Dana RA, Montrucchio L (1986) Dynamic complexity in duopoly games. J Econ Theory 40:40–56

Edgeworth FY (1897) La teoria pura del monopolio. Giornale degli Economisti 15:13–31

Guckenheimer J, Holmes P (1986) Nonlinear oscillations, dynamical systems, and bifurcations of vector fields. Springer, Berlin

Hotelling H (1929) Stability in competition. Econ J 39:41–57

Kuznetsov YA (1995) Elements of applied bifurcation theory. Springer, Berlin

Mira C, Gardini L, Barugola A, Cathala JC (1996) Chaotic dynamics in two-dimensional noninvertible maps. World Scientific, Singapore

Palander TF (1939) Konkurrens och marknadsjämvikt vid duopol och oligopol. Ekonomisk Tidskrift 41:124–145, 222–250

Puu T (1991) Chaos in duopoly pricing. Chaos, Solitons Fractals 1:573–581. Republished in Rosser JB (ed) Complexity in economics (Edward Elgar Publishing Inc., 2004)

Puu T (1996) Complex dynamics with three oligopolists. Chaos, Solitons Fractals 7:2075–2081

Puu T (1998) The chaotic duopolists revisited. J Econ Behav Organ 33:385–394

Puu T (2000, 2003) Attractors, bifurcations, & chaos - nonlinear phenomena in economics. Springer, Berlin. ISBN 3-540-66862-4. Second revised and enlarged edition (Springer-Verlag, ISBN 3-540-40226-8)

Puu T (2005) Complex oligopoly dynamics. In: Lines M (ed) Nonlinear dynamical systems in economics. CISM-Springer lecture notes in economics and mathematical systems, vol. 796. Springer, Berlin, pp 165–186

Rand D (1978) Exotic phenomena in games and duopoly models. J Math Econ 5:173–184

Robinson JV (1933) The economics of imperfect competition. Cambridge University Press, Cambridge

Tramontana F, Gardini L, Puu T (2010) Global bifurcations in a piecewise smooth Cournot duopoly. Chaos, Solitons Fractals 43:15–23

Chapter 4
Cournot II: Returns to Scale and Stability

Abstract Cournot did not just invent quantity competition duopoly; he considered the entire scale of steps from monopoly to perfect competition and described the proper rules for monopoly pricing and perfect competition pricing, assuming oligopoly price to be between these extremes. He considered the route in terms of adding new competitors to an already existent market, which, by the way, explained why he focused on quantity competition. Given the simple setups for Cournot duopoly—linear demand and constant marginal cost—it came as a shock when it was discovered that the models became unstable if the competitors exceeded some rather small number. What use is it to know that an increasing number of competitors has the perfect competitive state as target if both the asymptotic state and the route are destabilized? It occurred to the present author that the problem is on the side of production. Constant marginal costs, or constant returns to scale, means that we deal with firms of potentially infinite size. Any small profit margin between price and cost could be multiplied up any number of times through expanding production. Once this is realized, destabilization is no longer surprising. Further, it was probably not such a comparison Cournot was after; he might have wanted to compare a small number of **big** firms to a large number of **small** firms. But small and big cannot be defined without decreasing returns and capacity limits. Such can be introduced through the use of a nonstandard variant of the Constant Elasticity of Substitution production function. One just has to fix capital inputs through an act of investment, and production with the remaining variable inputs is automatically provided with a capacity limit for the lifetime of invested capital. This actually removes the destabilization problem. Once one is that far, it is natural to assume that capital wears out, and one may try to formulate an endogenous process for capital wear and regeneration, which is, in fact, done. However, many alternatives to this part of modelling are conceivable. The stability issue is bound to become complicated with capital renewal included, and there are interesting scenarios with, for instance, never ending oscillations and spontaneous formation of synchronized competitor groups.

T. Puu, *Disequilibrium Economics*, https://doi.org/10.1007/978-3-319-74415-5_4

4.1 Introduction

Cournot investigated the development of competition in an industry, from monopoly and duopoly to perfect competition. Any new industry is usually established through a few pioneering firms, and gradually starts to grow in terms of the number of competitors.

If we want to show how oligopoly can develop into a competitive economy, two conditions must hold: First, the Cournot equilibrium must seamlessly turn into the competitive equilibrium. Second, the Cournot equilibrium must not be destabilized through the addition of new competitors. The first condition does not present any problem, but the latter has been seriously questioned.

4.1.1 The "Theocharis Problem"

In the Anglo-Saxon world, Theocharis (1959) has been credited with the following problem. Assume we deal with a linear demand function and Cournot oligopolists facing constant marginal costs. Then the Cournot equilibrium point is stable with two competitors, and becomes unstable with four. In the intermediate case with three competitors, the equilibrium is neutrally stable and any initial oscillation remains for ever with constant amplitude. This is an obvious problem if we want a Cournot equilibrium to be transformed into the perfect competition case. Notably this "Theocharis Problem" was fully analyzed 20 years earlier by Palander (1936, 1939).

The argument is simple: Such a linear model produces reaction functions whose derivatives are $-\frac{1}{2}$. With n competitors there are then $n-1$ eigenvalues equal to $\frac{1}{2}$, associated with pairwise differences of the of the outputs, and one eigenvalue equal to $-\frac{n-1}{2}$, associated with the sum of the outputs. The last is the cause of the trouble. If $n > 3$, then $\left|-\frac{n-1}{2}\right| > 1$.

Neither Palander, nor Theocharis, considered the global dynamics. An unstable linear model like this is bound to result in negative outputs, so, to make sense, it must be constrained. If so, the model becomes piecewise linear, i.e., nonlinear. This was considered in Puu and Rúiz Marin (2006), and in more detail in Cánovas et al. (2008). It turned out that the resulting dynamics was very simple. The number of competitors remaining on the market could be stated in terms of relations among their marginal costs.

If one or two competitors remained, then a stable monopoly or duopoly was established. If more than three competitors remained, then the system would always go to a 2-period oscillation.

The Iso-Elastic Case

As the $-\frac{1}{2}$ slopes come from the marginal revenue and ultimately the inverse demand functions, one might think that this was due to the linear demand function, though the same happens with the model with iso-elastic demand discussed in Chap. 3 if we increase the number of competitors. The border of instability is only pushed to cases with five firms instead of four "pentopoly" or whatever we wish to call the case. This was shown by Ahmed and Agiza (1998).

4.1.2 Capacity Limits

The problem seems to arise, not from the demand function, but from the assumption of constant marginal costs. After all, what Palander and Theocharis as well as Ahmed and Agiza considered was the addition of competitors that operate under constant returns.

However, constant returns means assuming firms that are potentially infinitely large in capacity. It is then neither surprising nor interesting that the addition of competitors has this destabilizing effect.

Further, it is not the comparison we want to make. We want to compare the case of *many small firms* to the case of *few large firms*. This is impossible to represent unless we assume capacity limits as suggested already by Edgeworth (1897).

4.2 Duopoly with Capacity Limits

4.2.1 Production and Cost

In Chap. 2 we found that a convenient production function that incorporates capacity limits could be obtained using the CES function $q_i^\rho = k_i^\rho + l_i^\rho$, $\rho < 0$, with two inputs, capital k_i and labour l_i.

These denominations should not be taken too literally. Capital in the model represents everything durable in the production process, which is invested for some longer period, wears out during its service time, and must eventually be replaced if production is to be continued. Labour, on the other hand, represents any current inputs which can be accessed continuously. This is of some importance when input costs are considered, as they may vary considerably among firms. A conveniently simple case is obtained with exponent $\rho = -1$, so we assume this for both duopolists

$$q_1 = \frac{k_1 l_1}{k_1 + l_1} \tag{4.1}$$

$$q_2 = \frac{k_2 l_2}{k_2 + l_2} \tag{4.2}$$

In these functions both inputs become limitational. Output q_i can neither exceed the input of capital k_i nor that of labour l_i, no matter how one tries to compensate it with the other input. This may lead to a difference in capacity limits as capital is fixed during the respective investment periods.

We can solve (4.1)–(4.2) for l_i to see how much labour is needed to produce different quantities q_i, given the limiting capital stock k_i

$$l_1 = \frac{k_1 q_1}{k_1 - q_1} \tag{4.3}$$

$$l_2 = \frac{k_2 q_2}{k_2 - q_2} \tag{4.4}$$

Recall at this point that the denominators are positive as production cannot exceed capacity limits.

Denote capital cost (rent) r_i and labour cost (wage) w_i. As mentioned it is reasonable to assume that they can be very different among the competitors. Upon substitution from (4.3)–(4.4) the cost functions $C_i = r_i k_i + w_i l_i$ are

$$C_1 = r_1 k_1 + w_1 \frac{k_1 q_1}{k_1 - q_1} \tag{4.5}$$

$$C_2 = r_2 k_2 + w_2 \frac{k_2 q_2}{k_2 - q_2} \tag{4.6}$$

See Fig. 4.1 for illustration. Marginal costs then become

$$MC_1 = \frac{\partial C_1}{\partial q_1} = w_1 \frac{k_1^2}{(k_1 - q_1)^2} \tag{4.7}$$

$$MC_2 = \frac{\partial C_2}{\partial q_2} = w_2 \frac{k_2^2}{(k_2 - q_2)^2} \tag{4.8}$$

which are very convenient expressions as they are even squares, as was (almost) the marginal revenue obtained with iso-elastic demand.

4.2.2 *Long Run Cost*

Note that these are short run cost functions, given that capital is fixed during the investment period. However, once capital has worn out and has to be renewed, both capital and labour are fully variable. As the production functions (4.1)–(4.2) are

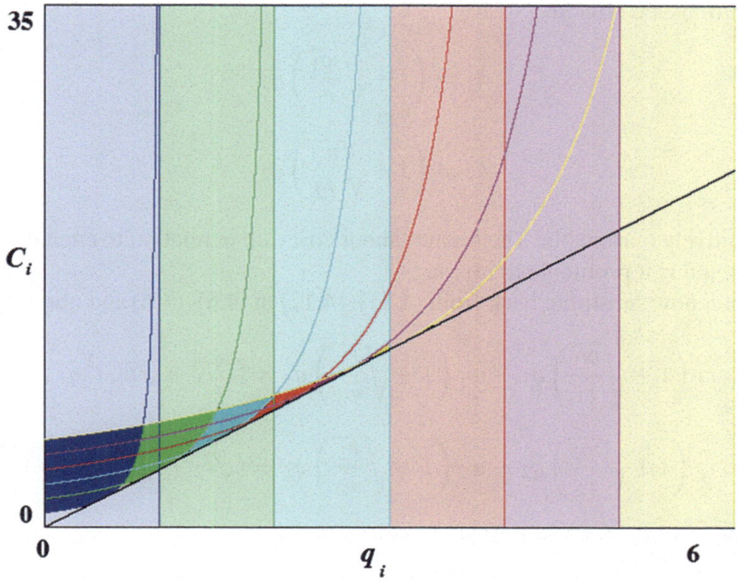

Fig. 4.1 Short run cost functions $C_i = rk_i + w\frac{k_i q_i}{k_i - q_i}$ for $k_i = 1, 2, \ldots 6$ and their envelope the long run cost function $C_i = \left(\sqrt{r} + \sqrt{w}\right)^2 q_i$. Note that each short run curve touches the long run curve in just one point where production is optimal for the capacity chosen. As a distinct set of curves the short run ones intersect, and form intervals in which that particular capacity is better than the next lower and the next higher

linear homogenous we are back at constant returns in the long run—it is the locking of one input that produces capacity limits. This is a point much stressed in general by Ragnar Frisch (1965).

The (long run) cost functions can then be obtained as usual, minimizing cost with respect to all inputs, using the production function as constraint. But it is instructive to start out from our already derived short term cost functions (4.5)–(4.6).

We just equate the derivatives of (4.5)–(4.6) with respect to k_i to zero and solve.

$$\frac{\partial C_1}{\partial k_1} = r_1 - w_1 \frac{q_1^2}{(k_1 - q_1)^2} = 0 \tag{4.9}$$

$$\frac{\partial C_2}{\partial k_2} = r_2 - w_2 \frac{q_2^2}{(k_2 - q_2)^2} = 0 \tag{4.10}$$

The solutions are obvious

$$k_1 = \left(1 + \sqrt{\frac{w_1}{r_1}}\right) q_1 \tag{4.11}$$

$$k_2 = \left(1 + \sqrt{\frac{w_2}{r_2}}\right) q_2 \tag{4.12}$$

and intuitively reasonable: The higher labour costs are in relation to capital cost, the more capital is it profitable to invest.

We can now substitute back from (4.11)–(4.12) in (4.5)–(4.6) and obtain

$$C_1 = r_1 \left(1 + \sqrt{\frac{w_1}{r_1}}\right) q_1 + w_1 \left(1 + \sqrt{\frac{r_1}{w_1}}\right) q_1 = \left(\sqrt{r_1} + \sqrt{w_1}\right)^2 q_1 \tag{4.13}$$

$$C_2 = r_2 \left(1 + \sqrt{\frac{w_2}{r_2}}\right) q_2 + w_2 \left(1 + \sqrt{\frac{r_2}{w_2}}\right) q_2 = \left(\sqrt{r_2} + \sqrt{w_2}\right)^2 q_2 \tag{4.14}$$

Please check this! Marginal revenues are now constant again and equal $\left(\sqrt{r_1} + \sqrt{w_1}\right)^2$, $\left(\sqrt{r_2} + \sqrt{w_2}\right)^2$. We claimed that the present derivation was instructive. Well, it is because it shows that the long run cost functions are envelopes to the families of short run cost functions, graphically touching the former in just one point each.

4.2.3 *Reaction Functions*

For obtaining the reaction functions, we again use the isoelastic demand function, i.e., $p = \frac{1}{q_1+q_2}$, as we recall derived from a Cobb-Douglas utility function.

Short run profits for the duopolists are then

$$V_1 = \frac{q_1}{q_1 + q_2} - r_1 k_1 - w_1 \frac{k_1 q_1}{k_1 - q_1} \tag{4.15}$$

$$V_2 = \frac{q_2}{q_1 + q_2} - r_2 k_2 - w_2 \frac{k_2 q_2}{k_2 - q_2} \tag{4.16}$$

Differentiating with respect to q_i and equating to zero

$$\frac{\partial V_1}{\partial q_1} = \frac{q_2}{(q_1 + q_2)^2} - w_1 \frac{k_1^2}{(k_1 - q_1)^2} = 0 \tag{4.17}$$

$$\frac{\partial V_2}{\partial q_2} = \frac{q_1}{(q_1 + q_2)^2} - w_2 \frac{k_2^2}{(k_2 - q_2)^2} = 0 \tag{4.18}$$

The full advantage of the combination of assumptions is now apparent; as q_i, w_i are nonnegative economic variables we can write the conditions

$$\frac{\sqrt{q_2}}{q_1 + q_2} = \sqrt{w_1}\frac{k_1}{k_1 - q_1} \tag{4.19}$$

$$\frac{\sqrt{q_1}}{q_1 + q_2} = \sqrt{w_2}\frac{k_2}{k_2 - q_2} \tag{4.20}$$

from which the reaction functions can easily be solved

$$q_1' = k_1 \frac{\sqrt{\frac{q_2}{w_1}} - q_2}{k_1 + \sqrt{\frac{q_2}{w_1}}} \tag{4.21}$$

$$q_2' = k_2 \frac{\sqrt{\frac{q_1}{w_2}} - q_1}{k_2 + \sqrt{\frac{q_1}{w_2}}} \tag{4.22}$$

Compared to the main case for reaction functions without capacity limits in Chap. 3, we note that the denominator tends to damp both the root term and the linear term thus promising more stability.

There are now four parameters to play with, capacities k_1, k_2 and wage rates w_1, w_2. Again, (4.21)–(4.22) return negative phase variables when $w_1 q_2 > 1$ or $w_2 q_1 > 1$, so it is advisable to use an adaptive format for the map.

$$q_1' = \begin{cases} (1 - \delta_1) q_1 + \delta_1 k_1 \frac{\sqrt{\frac{q_2}{w_1}} - q_2}{k_1 + \sqrt{\frac{q_2}{w_1}}} & q_2 < \frac{1}{w_1} \\ (1 - \delta_1) q_1 & q_2 \geq \frac{1}{w_1} \end{cases} \tag{4.23}$$

$$q_2' = \begin{cases} (1 - \delta_2) q_2 + \delta_2 k_2 \frac{\sqrt{\frac{q_1}{w_2}} - q_1}{k_2 + \sqrt{\frac{q_1}{w_2}}} & q_1 < \frac{1}{w_2} \\ (1 - \delta_2) q_2 & q_1 \geq \frac{1}{w_2} \end{cases} \tag{4.24}$$

Adaptation adds two more parameters δ_1, δ_2, so we have a total of six. For numerical study, wage rates and capacities seem to be of greater interest, so we can fix δ_1, δ_2 at some value. We also propose to take $k_1 = 1, w_1 = 1$ as a reference, though we have no reason to believe that our capacities and wage rates this time, like marginal costs in Chap. 3 influence the system only through their ratios. However, we only have two dimensions for bifurcation diagrams to play with.

In order to get the coordinates of the Cournot point, we have to solve (4.21)–(4.22), identifying $q_1' = q_1$ and $q_2' = q_2$, as a simultaneous system. The present author did not manage with the algebra, but the reader may give it another try.

4.2.4 Numerics

To find out something about the system (4.23)–(4.24), we can as usual use the computer. The first item is the bifurcation diagram in k_2, w_2 parameter plane. We keep $k_1 = 1, w_2 = 1$ fixed, and $\delta_1 = \delta_2 = 0.99$ as well. In Fig. 4.2 we focus a promising portion of the parameter plane, $1 < k_1 < 3$, and $7.5 < w_2 < 10$, and check out the 45 lowest periodicities. Obviously, it is the wage rate, i.e., the cost for variable inputs that is most responsible for instability, capacity having but minor influence.

As we see there are few low periods after the dark blue fixed point area at the bottom is left. Most periods we checked lie in the interval (20, 30). Yet there remain a lot of black bands which indicate truly high periodicities or even chaos. There are also dappled bands, which indicate coexistence of attractors of different periods.

To check out about chaos we complement the bifurcation diagram with one that displays areas of positive Lyapunov exponents. As we see in Fig. 4.3, the bands fit well into the black spaces of Fig. 4.2, and vice versa. The colour code for Fig. 4.3 is that blue indicates slightly positive Lyapunov exponent, whereas reddish to yellow indicate higher positivity.

Finally, in Fig. 4.4 we show a bifurcation series in the q_1, q_2 phase space as parameter w_2 increases through the interval (8, 10), displayed in Figs. 4.2 and 4.3.

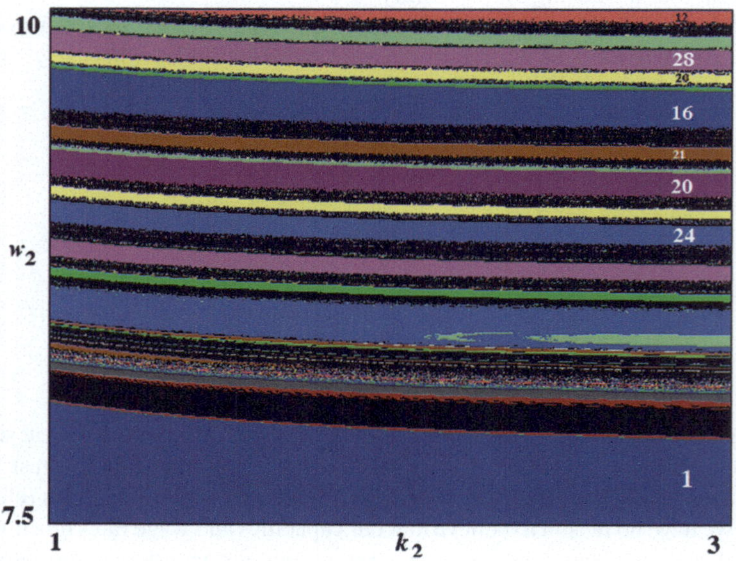

Fig. 4.2 The bifurcation plane k_2, w_2 (for the first competitor the parameters are normalized to $k_1 = 1, w_1 = 1$). As we see k_2 has little influence on the bifurcation structure, whereas minor change in w_2 leads to a cascade of bifurcations, mostly periodicities higher than 20. Remaining black bands indicate orbits of period higher than 45 or even chaos

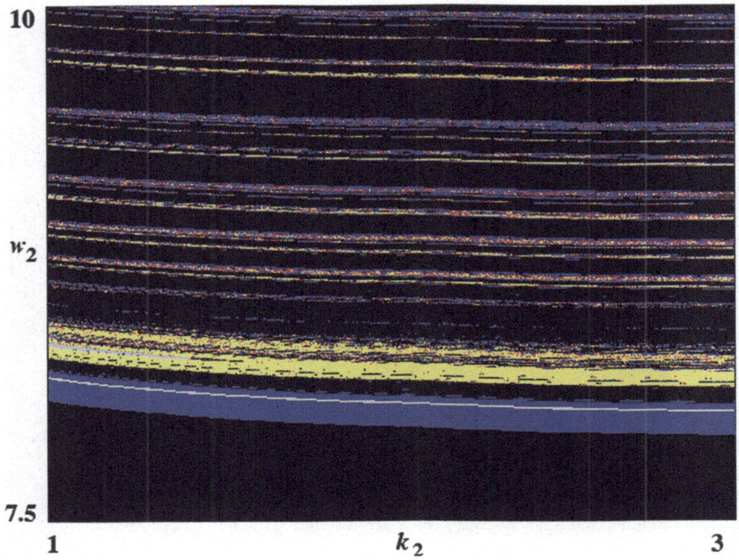

Fig. 4.3 As a complement to the previous picture we show bands of positive largest Lyapunov exponent. The colour code is from blue to yellow (mildly positive to higher positive). Compare to the previous picture to see how the colored bands fit into the black ones left in the previous picture

The first event is the destabilization of the fixed point to an invariant curve. As we model in discrete time, it is not a continuous closed curve, which also means that contiguous points do not follow neighbour to neighbour. What appears as a curve is generated by repeated showers of points at different places of the curve, only eventually generating the image of a closed curve.

We see one more green version of such an invariant curve. Then comes the light blue periodic orbit of six large dots, which eventually bifurcates to the dark blue periodic orbit of much higher period. Finally, there is the yellow chaotic looking set of four irregular shapes. Note that, the displayed attractors are only a selection of cases between which we may see innumerable other attractors. It is instructive to run the map (4.23)–(4.24) like this, though for pictures we need a very small selection of cases, otherwise everything becomes just cluttered.

4.3 Increasing Competition

Reconsider now the issue Cournot focused through adding more and more competitors to a duopoly market, though now assuming production under capacity constraints.

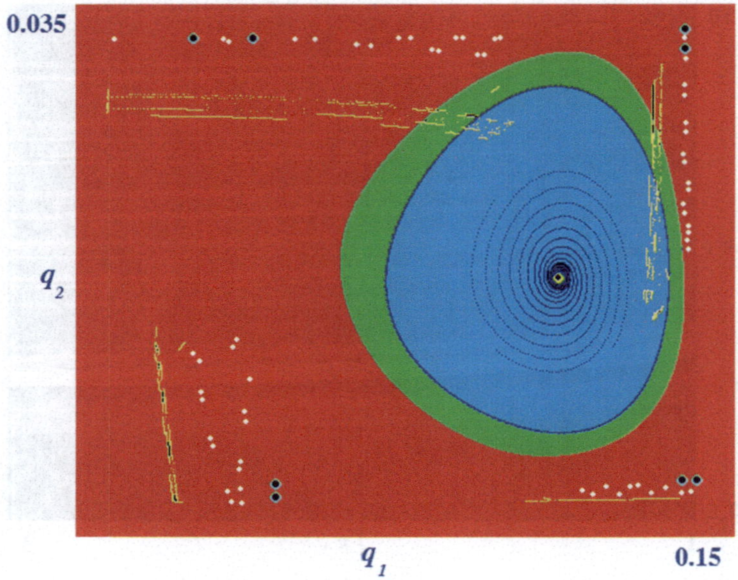

Fig. 4.4 A bifurcation sequence in terms of snapshots of the phase plane. We see how the Cournot equilibrium point loses stability, and how the blue orbit spirals out to a closed invariant curve. Note again that this curve is not anything like a limit circle in a continuous dynamic system. It is not a continuum, and it is traced through showers of distinct points at different parts of the orbit, which eventually create the illusion of a curve. One larger green closed orbit is shown, but after a while it breaks up in periodic orbits. We see two samples of this: An eight period orbit shown by large dots, and further an orbit of much higher period depicted with small white dots. Eventually the yellow attractor has an apparently fractal (chaotic?) appearance

To be quite true, this at any moment holds for some of the firms, maybe the majority, whereas the rest are renewing capital stock and hence face constant returns with both capital and labour inputs variable.

Suppose there are a total of n firms. Their total supply is now

$$Q = \sum_{i=1}^{i=n} q_i \tag{4.25}$$

so, assuming the isoelastic market demand, price becomes

$$p = \frac{1}{Q} \tag{4.26}$$

It is convenient for succinct formulas to define residual supply

$$Q_i = Q - q_i \tag{4.27}$$

i.e., the supply not under the control of a particular firm.

Thereby (4.25) can be written

$$p = \frac{1}{Q_i + q_i} \tag{4.28}$$

In order to simplify, the numerator has been normalized to unity. It represents the sum of the budget shares that all the consumers spend on the particular commodity, but, we have two degrees of freedom, which we can use to get rid of any superfluous parameter.

4.3.1 Production and Cost Functions

As for production functions we only need to recapitulate (4.1)–(4.2) for the general firm operating under capacity constraint

$$q_i = \frac{k_i l_1}{k_i + l_i} \tag{4.29}$$

which works both if k_i, l_i can be varied in periods of re-investment, or only l_i. In the latter case we again solve (4.29) for

$$l_i = \frac{k_i q_1}{k_i - q_i} \tag{4.30}$$

which substituted in the cost function gives

$$C_i = r_i k_i + w_i \frac{k_i q_i}{k_i - q_i} \tag{4.31}$$

as in (4.5)–(4.6).

The long run cost is then, through minimizing (4.31) with respect to k_i and substituting back obtained as

$$C_i = \left(\sqrt{r_i} + \sqrt{w_i}\right)^2 q_i \tag{4.32}$$

There is no point in repeating this derivation anew.

4.3.2 Reaction Functions

From (4.31) and (4.32) we get the short run marginal costs

$$MC_i = w_i \frac{k_i^2}{(k_i - q_i)^2} \tag{4.33}$$

quite as in (4.7)–(4.8).

For the long run likewise

$$MC_i = \left(\sqrt{r_i} + \sqrt{w_i}\right)^2 \tag{4.34}$$

We do not complicate things through using different symbols for short and long run costs—there is little risk of confusion.

Using our new symbols for residual demand, marginal revenue becomes

$$MR_i = \frac{\partial}{\partial q_i}\left(\frac{q_i}{Q_i + q_i}\right) = \frac{Q_i}{(Q_i + q_i)^2} \tag{4.35}$$

which, of course, is relevant in short as in long run action.

Equating marginal revenue (4.35) to marginal cost (4.33)

$$\frac{Q_i}{(Q_i + q_i)^2} = w_i \frac{k_i^2}{(k_i - q_i)^2} \tag{4.36}$$

is again easily solved for the reaction functions

$$q_i' = k_i \frac{\sqrt{\frac{Q_i}{w_i}} - Q_i}{k_i + \sqrt{\frac{Q_i}{w_i}}} \tag{4.37}$$

The best reply now obviously depends on all the delayed moves of the other competitors, more precisely their sum.

As usual (4.37) returns a negative best reply if $w_i Q_i > 1$, so we have to constrain the formula to the interval $w_i Q_i \leq 1$, and if we want to iterate the system, we should again choose the adaptive variant

$$q_i' = \begin{cases} (1 - \delta_i)\, q_i + \delta_i k_i \frac{\sqrt{\frac{Q_i}{w_i}} - Q_i}{k_i + \sqrt{\frac{Q_i}{w_i}}} & Q_i \leq \frac{1}{w_i} \\ (1 - \delta_i)\, q_i & Q_i > \frac{1}{w_i} \end{cases} \tag{4.38}$$

By definition

$$Q_i = \sum_{j=1}^{j=n} q_j - q_i \tag{4.39}$$

We also need the reaction functions for long run adjustment. To this end equate marginal revenue (4.35) to marginal cost (4.34)

$$\frac{Q_i}{(Q_i + q_i)^2} = \left(\sqrt{r_i} + \sqrt{w_i}\right)^2 \tag{4.40}$$

The reaction function is then

$$q_i' = \frac{\sqrt{Q_i}}{\sqrt{r_i} + \sqrt{w_i}} - Q_i \tag{4.41}$$

which the reader easily recognizes as the main case from Chap. 3.

Again, the iteration should take the negative returns in consideration and choose the adaptive version

$$q_i' = \begin{cases} (1-\delta_i)\, q_i + \delta_i \left(\frac{\sqrt{Q_i}}{\sqrt{r_i}+\sqrt{w_i}} - Q_i\right) & Q_i \le \frac{1}{\left(\sqrt{r_i}+\sqrt{w_i}\right)^2} \\ (1-\delta_i)\, q_i & Q_i > \frac{1}{\left(\sqrt{r_i}+\sqrt{w_i}\right)^2} \end{cases} \tag{4.42}$$

Note that the clause for the main branch is stronger in (4.42) than in (4.38) as $\left(\sqrt{r_i} + \sqrt{w_i}\right)^2 > \sqrt{w_i}$. This is as expected because in the long run the capital cost as well is taken in account.

4.4 Equilibrium and Its Stability

4.4.1 Cournot Equilibrium

The question is how we should define Cournot equilibrium in this polypoly case. Obviously it can only be done for the long run map (4.41). It is obvious that if things change every now and then, different firms always reinvesting and changing capital stock due to market conditions, we could not speak of any kind of equilibrium.

So identifying $q_i' = q_i$ and rearranging the reaction function, we get

$$q_i + Q_i = \frac{\sqrt{Q_i}}{\sqrt{r_i} + \sqrt{w_i}} \tag{4.43}$$

or, defining,

$$c_i = \left(\sqrt{r_i} + \sqrt{w_i}\right)^2 \tag{4.44}$$

$$Q = \sqrt{\frac{Q_i}{c_i}} \tag{4.45}$$

as $q_i + Q_i = Q$ by definition. Taking squares, solving for

$$Q_i = c_i Q^2 \tag{4.46}$$

and taking the sum, we get

$$\sum_{i=1}^{i=n} Q_i = Q^2 \sum_{i=1}^{i=n} c_i \tag{4.47}$$

However, as $Q_i = Q - q_i$, we also get

$$\sum_{i=1}^{i=n} Q_i = (n-1)\, Q \tag{4.48}$$

Identifying the equations, and dividing by Q, we can solve for

$$Q = \frac{n-1}{n}\frac{1}{c} \tag{4.49}$$

where

$$c = \frac{1}{n}\sum_{i=1}^{i=n} c_i \tag{4.50}$$

is average marginal cost for all the competitors.

Recalling our isoelastic demand function $p = \frac{1}{Q}$, we find that $p = \frac{n}{n-1}c$, which makes sense as only for an infinite number of competitors we get marginal cost pricing $p = c$, whereas for a monopoly price goes to infinity, an anomaly of the iso-elastic function as we already know. In the series of steps towards perfect competition, for duopoly price is twice the marginal cost $p = 2c$. This was what Cournot was reasoning about even if he did not use the isoelastic function.

Once we have total market demand we can easily obtain the supply for each firm in the Cournot point. From (4.45) $Q_i = c_i Q^2$, so $q_i = Q - Q_i = Q\,(1 - c_i Q)$ or

$$q_i = \frac{n-1}{n}\frac{1}{c}\left(1 - \frac{c_i}{c}\frac{n-1}{n}\right) \tag{4.51}$$

Should the firms be identical, $c_i = c$, then $q_i = \frac{n-1}{n^2}\frac{1}{c}$, otherwise the firms that face higher production cost contribute less to the total supply.

Concerning capital (capacity) we can also calculate it for the firms in Cournot equilibrium. Recall that we are considering the long run case, so capital must be considered as calculated at the optimal scale, i.e., at the point where the short run cost curve touches its long run envelope. The formula vas given in (4.11) in the duopoly section and we do not repeat it again. The result, using (4.51) and definition (4.44) is

$$k_i = \sqrt{\frac{c_i}{r_i}\frac{n-1}{n}\frac{1}{c}\left(1 - \frac{c_i}{c}\frac{n-1}{n}\right)} \tag{4.52}$$

As capital is optimal for each firm, one needs not be concerned with which firms invest and which do not, because nobody will change capital from the optimal attained.

4.4.2 *Cost Structure in Cournot Equilibrium*

We add a little digression about the cost structure in the Cournot point. Obviously, with an increasing number of competitors, the admissible dispersion of production costs among firms in Cournot equilibrium becomes smaller. In the limit as n goes to infinity, perfect competition, all the firms have to be identical. Increasing competition thus requires more and more homogeneity among the firms. This seems to be quite an interesting conclusion.

We calculated the outputs in the Cournot point for a duopoly (3.6)–(3.7): $q_1 = \frac{c_2}{(c_1+c_2)^2}$, $q_2 = \frac{c_1}{(c_1+c_2)^2}$. However, it is worth noting that with three or more competitors, things are different, we get constraints on the possible cost distribution among the firms.

From (4.51) q_i becomes negative if

$$c_i \leq \frac{n}{n-1}c \tag{4.53}$$

is not fulfilled, i.e., if the marginal cost for the i:th firm exceeds the fraction $\frac{n}{n-1}$ of the average on the market; twice for duopoly three halves for triopoly, etc. As c_i is also in the r.h.s. definition of c, we might prefer a formula where c_i compares only with the other c_j, $j \neq i$. To this end define

$$C = \sum_{i=1}^{i=n} c_i \tag{4.54}$$

which means

$$nc = C \tag{4.55}$$

and

$$C_i = C - c_i \tag{4.56}$$

Using (4.54)–(4.56) in (4.53) one gets $c_i < \frac{n}{n-1}\frac{C_i+c_i}{n}$. Moving c_i to the l.h.s. we finally get

$$c_i \leq \frac{n-1}{n-2}\frac{C_i}{n-1} \tag{4.57}$$

The ratio $\frac{C_i}{n-1}$ is the average of the marginal costs for all competitors except the i:th. The condition thus states that the marginal cost for each and every firm that remains in Cournot equilibrium must be less than the multiplier $\frac{n-1}{n-2}$ of the average of those of the others.

For a triopoly this multiplier equals 2, for a quadropoly (tesseropoly?) 1.5 etc. In the limit the multiplier goes to 1. Hence with few competitors there is still a possible variation in the cost spectrum.

However, consider the competitive case. Then each marginal cost must be less than or equal to the average for the others, which means that all marginal costs have to be equal. It seems this has never been investigated. Check for instance what the maximum dispersion in the cost spectrum may be when $n = 3, 4, 5$. It is instructive and not quite trivial.[1]

4.4.3 Stability of Cournot Equilibrium

Consider now again the map (4.42). For discussing stability we need to calculate its derivatives. Suppose the competitors are numerous enough to assume that the firms that remain active in the Cournot equilibrium are identical, i.e., $r_i = r$ and $w_i = w$. Further, use the shorthand for cost $\sqrt{r} + \sqrt{w} = \sqrt{c}$. Obviously we are in the main

[1] From (4.53) we can calculate the relative standard deviation

$$\frac{1}{n}\sum_{i=1}^{i=n}\left(\frac{c_i}{c} - 1\right)^2 = \frac{1}{(n-1)^2}$$

which no doubt goes to zero as n approaches infinity.

branch when we consider Cournot equilibrium, so

$$q_i' = (1 - \delta_i)\, q_i + \delta_i \left(\sqrt{\frac{Q_i}{c}} - Q_i \right) \tag{4.58}$$

Note that complementary demand $Q_i = \sum_{j \neq i} q_j$ is the sum of outputs for all the other competitors. Hence the derivatives are

$$\frac{\partial}{\partial q_i} q_i' = (1 - \delta_i) \tag{4.59}$$

and

$$\frac{\partial}{\partial q_j} q_i' = \delta_i \left(\frac{1}{2} \sqrt{\frac{1}{c Q_i}} - 1 \right) \tag{4.60}$$

As we consider stability at the Cournot point, we substitute for its coordinates from (4.46) and (4.49). Thus

$$\frac{\partial}{\partial q_j} q_i' = -\frac{1}{2} \delta_i \left(\frac{n-2}{n-1} \right) \tag{4.61}$$

The stability matrix thus becomes

$$\mathbf{M} = \begin{bmatrix} 1 - \delta_1 & \cdots & -\frac{1}{2}\delta_1 \left(\frac{n-2}{n-1}\right) \\ \cdot & \cdot & \cdot \\ \cdot & \cdot & \cdot \\ \cdot & \cdot & \cdot \\ -\frac{1}{2}\delta_n \left(\frac{n-2}{n-1}\right) & \cdots & 1 - \delta_n \end{bmatrix} \tag{4.62}$$

The eigenvalues then become

$$\lambda_1, \ldots \lambda_{n-1} = 1 - \frac{1}{2} \delta \frac{n}{n-1}, \qquad \lambda_n = 1 - \delta \frac{n}{2} \tag{4.63}$$

As the adaptation coefficient $0 < \delta < 1$, it is obvious that all eigenvalues but one are in the unit interval, $0 < \lambda_i < 1, i = 1, \ldots n - 1$. The last eigenvalue can, however, cause instability; $\lambda_n < -1$ whenever $\delta n > 4$. With no adaptation, $\delta = 1$, the limit is $n = 4$, quite as shown by Ahmed and Agiza. With $\delta = \frac{1}{2}$, we can have up to seven firms using long term adjustment in stable Cournot equilibrium. Clearly, if we let

$\delta \to 0$, we can have any number of competitors, and yet the equilibrium remains stable. In the limit $\lambda_n = -1$, and they are in neutral stability. This is not surprising as nothing happens in the system with such extreme adaptation.

4.5 Capital and Investment

4.5.1 Equilibrium with Identical Firms

To really focus Cournot's problem of industry structure and profit from our model that incorporates capacity constraints, we need to consider how production capacity, i.e., capital, is distributed between the firms, and, preferably also how it evolves due to internal forces in a dynamic process.

Say that there is a fixed total capacity K divided between the firms

$$K = \sum_{i=1}^{i=n} k_i \tag{4.64}$$

From (4.38) recover the main branch of the short run iterative map

$$q_i' = k_i \frac{\sqrt{\frac{Q_i}{w_i}} - Q_i}{k_i + \sqrt{\frac{Q_i}{w_i}}} \tag{4.65}$$

It is easiest to consider the special case of identical firms where all $k_i = \frac{K}{n}$ and further $w_i = w$. For calculating the Cournot point coordinates, as always, identify $q_i' = q_i$. Hence

$$q_i = \frac{K}{n} \frac{\sqrt{\frac{Q_i}{w}} - Q_i}{\frac{K}{n} + \sqrt{\frac{Q_i}{w}}} \tag{4.66}$$

It is not too daring to use the intuition that $q_i = q$ for all firms, and further that $q = \frac{1}{n} Q$ and $Q_i = \frac{n-1}{n} Q$.

Hence

$$\frac{Q}{K} = \frac{\sqrt{\frac{n-1}{n}\frac{Q}{w}} - \frac{n-1}{n} Q}{\frac{K}{n} + \sqrt{\frac{n-1}{n}\frac{Q}{w}}} \tag{4.67}$$

which provides a quadratic equation for Q, total output in the Cournot equilibrium, given total capacity K, the wage rate w, and the number of competitors n:

$$Q^2 - K\left(2 + \frac{n}{n-1}wK\right)Q + K^2 = 0 \tag{4.68}$$

Choosing the correct branch, the solution reads

$$Q = K + \frac{1}{2}\frac{n}{n-1}wK^2 - \frac{1}{2}K\sqrt{4\frac{n}{n-1}wK + \left(\frac{n}{n-1}\right)^2 w^2K^2} \tag{4.69}$$

There are some interesting features of (4.69). (1) We have $Q < K$, which is good as the capacity limits must hold for each firm, and hence also for the entire branch. (2) The capacity utilization $\frac{Q}{K}$ increases with the number of competitors n. (3) Yet it approaches a limit which may be far below total capacity utilization,

$$\lim_{n\to\infty}\frac{Q}{K} = 1 - \frac{1}{2}wK\left(\sqrt{1+\frac{4}{wK}} - 1\right) \tag{4.70}$$

As we know the Cournot coordinates, we can now evaluate the stability matrix for the iteration, and see how its eigenvalues depend on the number of competitors.

Yet, this digression takes its departure from Cournot's original idea to compare markets with different numbers of competitors, few large to many small in our setting with capacity limits.

More interesting may be to design dynamical processes where new competitors enter in succession. There is then no reason to believe that they enter by sharing the total capacity with those already in the market; rather they would increase the total capacity.

4.5.2 *Some Numerics*

Let us try to formulate a dynamic process where the firms enter one by one through setting up a capital stock that is optimal, given the market situation at the moment. For simplicity we take capital rent r and wage rate w as the same for all firms and also constant over time.[2]

[2]Of course, we could let capital and labour costs change over time, thereby providing a driving force for evolution. However, taking such development as exogenously given tends to be a bit trivial. One could also incorporate capital and labour markets in the model to provide for internal forces for change, though one must then care to avoid assumptions which just make a model too messy.

This means that how much capital a firm needs only depends on how much it is optimal to produce at the moment when the firm invests in new capital. Now, this is not done every period as any capital installment is costly. As capital has a lifetime, and depreciates every period, a reasonable first assumption is that it has a fixed lifetime during which it provides for full service capacity, and eventually must be renewed after this has elapsed to enable production to be continued.[3]

Each time a firm invests it selects capital at the optimal scale to the output as stated in (4.11)–(4.12) for the duopoly case

$$k_i' = \left(1 + \sqrt{\frac{w}{r}}\right) q_i \tag{4.71}$$

As for q_i two results were recorded (4.42)

$$q_i' = \begin{cases} (1-\delta_i)\, q_i + \delta_i \left(\frac{\sqrt{Q_i}}{\sqrt{c}} - Q_i\right) & Q_i \le \frac{1}{c} \\ (1-\delta_i)\, q_i & Q_i > \frac{1}{c} \end{cases} \tag{4.72}$$

when the i:th firm reinvests, and (4.24)

$$q_i' = \begin{cases} (1-\delta)\, q_i + \delta k_i \frac{\sqrt{\frac{Q_i}{w}} - Q_i}{k_i + \sqrt{\frac{Q_i}{w}}} & Q_i < \frac{1}{w} \\ (1-\delta)\, q_i & Q_i \ge \frac{1}{w} \end{cases} \tag{4.73}$$

when it does not. When (4.72) applies then (4.71) is used, if not then $k_i' = k_i$.

As $c = \left(\sqrt{r} + \sqrt{w}\right)^2 > w$, we obviously have the clear intervals $Q_i < \frac{1}{c}$, $\frac{1}{c} \le Q_i < \frac{1}{w}$, and $\frac{1}{w} \le Q_i$ for the branches.

What is still missing is a condition for deciding when a firm invests, and when not. To this end define an indicator function

$$\sigma = t - ni \bmod T \tag{4.74}$$

where t is the current time period (advanced by $t' = t + 1$ at each iteration), n is the total number of competing firms, i the identification number of each firm, and T is the lifetime of each capital installment.

The i:th firm (re)invests whenever $\sigma = 0$, otherwise not. For cooking up a program we finally need to settle with what intervals the n firms enter into the market—it would be inappropriate to let them enter all at once, as they thereby always (re)invest simultaneously, which would contribute unnecessarily much to instability. So, specify this separation of first entry in terms of periods m, and we are done.

[3] Other assumptions about depreciation are, of course, equally reasonable, but fixed lifetime and sudden death is the simplest.

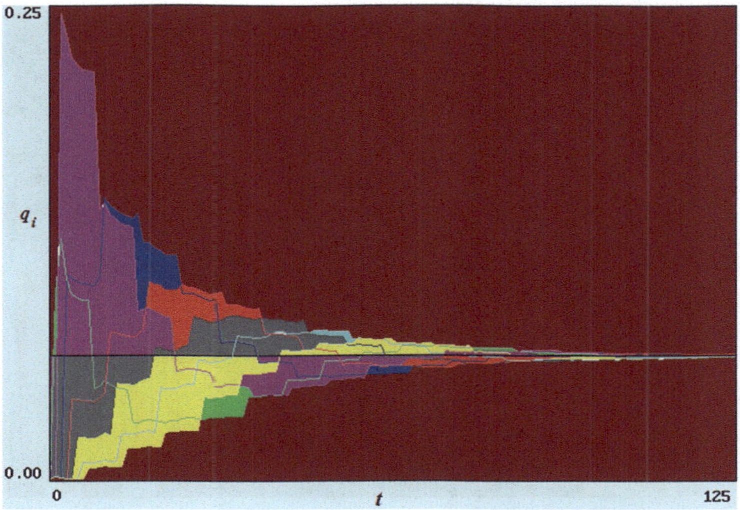

Fig. 4.5 Time traces for 14 firms that enter with intervals of 5 periods. To show the curves more clearly they are provided by filled support areas of similar shade, and we chose to overwrite firms earlier in business by those entering later. Therefore some of the firms are hardly visible close to equilibrium convergence. The order of entry in terms of colour is blue, green, brown, violet, ...yellow, and gray. The firms enter as mentioned with intervals of 5 periods. Capital wears out each 7 periods and has then to be replaced, though the quantity invested changes due to market conditions at the moment. Hence the big jumps in the supply curves. The minor irregularity of "skylines" represents periodic change while capital is fixed. As we see, the case with 14 competitors converges to a stable Cournot equilibrium, where the firms also become identical in terms of invested capital

We show two pictures from a simulation program, Figs. 4.5 and 4.6, one convergent to Cournot equilibrium, and one divergent. For all we choose a lifetime $T = 7$ periods for capital, and a separation for first entry of $m = 5$ periods. In the convergent case there are $n = 14$ firms, in the divergent $n = 16$ firms. A larger number of firms contributes to instability, but also to a tendency for synchronization of investments.

4.5.3 Endogenous Investment Decisions

Even if we now have a dynamic process where firms enter the market, invest capital and compete with the others, it is still all fairly mechanical—capital wears out after a certain number of periods, and has to be replaced if production is to go on.

In reality, a firm can choose to continue using equipment that perhaps should be scrapped, or it may replace capital with newer even before it is worn out. An idea used by the present author and collaborators, Dr Panchuk and Professor Cánovas,

Fig. 4.6 The same case as in the preceding illustration, though the number of competitors is now 16 instead of 14. Initially the firms still enter with intervals of 5 periods, and capital has a lifetime of 7 periods. As we see, the system after a first tendency to convergence diverges again, and eventually settles to a complex periodic structure. Such divergence may occur after many periods of orbits almost settled to equilibrium. Therefore the visual evidence is never conclusive. Increasing the number of competitors always tends to lead to instability, though things mainly depend on how many firms reinvest at the same time, as it is long term adjustment (under constant returns) that creates instability

is that replacing equipment should be an endogenous decision due to technical and economic circumstances. To this end one may consider continuous wear of capital and its remaining lifetime as variable.

The idea is also that an optimal use of capital would be with a production volume for which this capital stock is optimal, i.e., production at optimal scale as Frisch would have put it, $k_i = \sqrt{\frac{c}{r}} q_i$ from (4.71), or

$$q_i = \sqrt{\frac{r}{c}} k_i \tag{4.75}$$

If $\sqrt{\frac{r}{c}} k_i > q_i$, production is performed under full capacity, and one may assume that the remaining service lifetime is shortened by less than one full period. If, on the contrary, $\sqrt{\frac{r}{c}} k_i < q_i$, then capital wear reduces remaining lifetime by more than one period.One way to put this is as follows.

Assume a normal lifetime of capital in T periods. We suggest that during any current period it is changed to

$$T_i' = T_i - \kappa^{\sqrt{\frac{r}{c}} k_i - q_i} \tag{4.76}$$

where $\kappa > 0$. According to (4.75) $\sqrt{\frac{r}{c}}k_i$ is the optimal scale of operation. Hence T_i is reduced during an elapsed period by one if $\sqrt{\frac{r}{c}}k_i = q_i$, more than one the exponent is positive, and less than one if it is negative. In no case will however T_i be increased as long as $\kappa > 0$ holds.

Various assumptions are possible for a decision about when to renew capital. One is to assume that reinvestment occurs whenever the remaining lifetime is less than one period, $T_i < 1$. The rest of the bits and pieces have already been introduced, adding (4.76) and the criterion $T_i < 1$, enables us to formulate this version of the map.

We display one picture Fig. 4.7 using such a map, where again the basic capital duration period is 7, and there are 14 competitors. The fixed parameter stating the interval between first entry of the firms is now deleted. It is no longer needed as the reinvestment decision is internalized. Instead we assume that there is a random initial distribution of supplies and capacities. All will be internalized in due course of time.

To let the case be close to the previous, we chose a very high $\kappa = 500$, which makes the model come close to the rigid case where remaining lifetime is fixed.

The main scenario seems to be groupwise clustering and synchronization, a phenomenon calling for much more investigation.

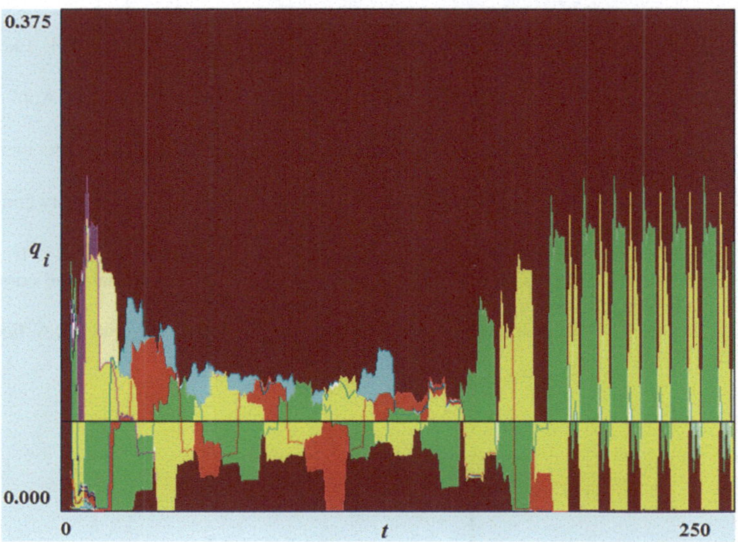

Fig. 4.7 Orbits for 14 firms in the case of flexible investment period. As we see, clustering in groups is the main scenario. There are at least five shades clearly visible, in this picture of 14 orbits. The blue, violet, and eventually brown shades disappear and only the yellow and green persist. This means that the other shades are overwritten by the last colour in each group. Except for spontaneous group-wise synchronisation, it seems that the groups alternate over time in high and low production

4.5.4 The Cournot Agenda

However, this Cournot agenda, as we may call it, is by no means exhausted by the attempts exposed in this Chapter, much interesting research remains to be done. Curiously, the full Cournot agenda seems to have interested researchers very little.

References

Ahmed E, Agiza NH (1998) Dynamics of a Cournot game with n competitors. Chaos, Solitons Fractals 10:1179–1184

Cánovas J, Puu T, Ruiz M (2008) The Cournot–Theocharis problem revisited. Chaos, Solitons Fractals 37:1025–1039

Cánovas J, Panchuk A, Puu T (2015) Asymptotic dynamics of a piecewise smooth map modelling a competitive market. Math Comput Simul 117:20–38

Edgeworth FY (1897) La teoria pura del monopolio. Giornale degli Economisti 15:13–31

Frisch R (1965) Theory of production. D. Reidel Publishing Company, Dordrecht

Heathfield FH, Wibe S (1987) An introduction to cost and production functions. MacMillan, London

Johansen L (1972) Production functions. North-Holland, Amsterdam

Palander TF (1936) Instability in competition between two sellers. In: Abstracts of papers presented at the research conference on economics and statistics held by the Cowles Commission at Colorado College, Colorado College Publications, General Series No. 208, Studies Series No. 21

Palander TF (1939) Konkurrens och marknadsjämvikt vid duopol och oligopol. Ekonomisk Tidskrift 41:124–145, 222–250

Puu T (2007) Layout of a new industry: from oligopoly to competition. Pure Math Appl 16:475–492

Puu T (2008) On the stability of Cournot equilibrium when the number of competitors increases. J Econ Behav Organ 66:445–456

Puu T, Norin A (2003) Cournot duopoly when the competitors operate under capacity constraints. Chaos, Solitons Fractals 18:577–592

Puu T, Panchuk A (2009) Oligopoly and stability. Chaos, Solitons Fractals 41:2505–2516

Puu T, Rúiz Marin M (2006) The dynamics of a triopoly Cournot game when the competitors operate under capacity constraints. Chaos, Solitons Fractals 28:403–413

Theocharis RD (1959) On the stability of the Cournot solution on the oligopoly. Rev Econ Stud 27:133–134

Chapter 5
Bertrand

Abstract Once oligopoly theory became an interesting target for researchers looking for application of nonlinear dynamics in economics, the supply of Cournot models almost exploded. Not so for the Bertrand case. It is important to realize that the difference does not lie in whether price or quantity is used for optimizing—both work equally well in all monopolistic markets. The difference lies in whether the competing commodities are considered as identical or as slightly different substitutes by the consumers. They cannot be both at the same time! Bertrand modelling was hampered because economists never supplied a convincing demand theory for close substitutes, at least not any precise model that might work in a global setting such as a dynamic process wandering around in an extended phase space. There, however, is one possibility to model substitutes, once proposed by Kelvin Lancaster. In his "new theory of demand" it is not the marketed commodities, but their "properties" as measurable performance scores, that enter the utility function. A marketed commodity is thus regarded as a bundle of such property scores, so the consumer can even combine different marketed commodities to obtain a desired property mix. The property score vector of each commodity is designed by the producer. For close substitutes the vectors are similar but not identical. This makes it possible to model Bertrand price competition using precise demand functions, which, of course, change whenever a design is changed. Unfortunately, the model setup does not permit solving for explicit reaction functions, which we need for cooking up a computer experiment program. Fortunately, it is possible to define close approximations to the true reaction functions—an unusual strategy, but it works. We find interesting dynamics scenarios even with this model. A particular advantage of using the Lancaster approach is that we can also consider choice of design as a means of competition along with pricing. As the Lancasterian properties get well defined shadow prices within the model, the value of each design can be quantified, and the competitors can choose a change whenever the value of the best possible design exceeds that of the current design by a sufficient amount. In this way we can see whether the system converges to identical designs or if the competitors rather seek very different characteristics of their products. Other solutions, such as periodic, may be possible, which may explain cyclic recurrence of designs. This study is not carried out in the book, but left as a stub to the reader. There is also another related stub. Recently the present author saw a few contributions trying to

T. Puu, *Disequilibrium Economics*, https://doi.org/10.1007/978-3-319-74415-5_5

combine Cournot and Bertrand by letting the competitors use both price and quantity adjustments. These simply do not work as **a commodity either is homogenous, or it is not**. There is one possibility though to put sense in this wish to combine, and that is to assume that there are different groups of consumers, some who regard the commodities as identical, some who do not. In this case both competitors might use price discrimination and sell in the different markets. A model, again based on Lancaster is devised as a stub, but analysis is left to the reader.

5.1 Some History

5.1.1 *Biographical*

Joseph Bertrand (1822–1900) was a French mathematician. His main fields were number theory and probability, though he seems not to have made any important contributions. He is known for a conjecture about prime numbers, and for translating Gauss's work on probability to French. He seems to have concentrated on administration as secretary in the French Academy of Sciences and in the Collège de France. This explains that he could publish his very short and aggressive denunciation of mathematical economics, notably the contributions by Augustin Cournot and Léon Walras in the prestigious Journal des Savants.

One sometimes sees reference to the Bertrand "model", as if he had produced an alternative to Cournot, but this is not true.[1] There is an important point in his attack concerning the Cournot model. However, Schumpeter in his great history of economic analysis, notes that the argument was so confused that nobody would have paid attention to if it had not been cleared up by Edgeworth a few years later. It is, by the way, notable that Bertrand's attack came half a Century after Cournot's work, so the aggressiveness can hardly be explained by any personal enmity as the men belonged to different generations.

5.1.2 *Bertrand's Argument*

The essence of Bertrand's critique of Cournot's duopoly model concerns the assumption that the oligopolists compete on a market for a homogenous good as seen by the consumers. The sellers compete through supply, and a uniform price is then established in accordance with the inverse demand function. Bertrand argues that if this is true, then any competitor could take the entire market through a but

[1] A bad habit is spreading among researchers of not checking original sources. The present author in his student days learned to never cite a reference he had not read. These days such precautions are forgotten. It is rather customary to cite references one has only heard of, which is a dangerous habit.

infinitesimal price cut. This, of course, would set up a series of successive price cuts. It is not clear where this process would end, but the argument demonstrates that the Cournot equilibrium could not be stable.

Harold Hotelling in 1929 constructed a unique case where the commodity is indeed physically homogeneous, though not indifferent to the consumers. This was achieved through placing the competitors at a geographical distance so that the costs of transportation would have to be added to the mill prices, which could now be different. This resulted in local monopoly areas with competition at the boundary.[2] In the sequel we will study Hotelling in an entire chapter.

Alternatively, the commodities supplied by different competitors might be considered as slightly different by the consumers, though essentially satisfying the same needs. This would provide for possibilities to charge different prices, and it would not be the case that all consumers would choose another supplier if there was an infinitesimal price cut.

The problem has been that though there are very precise local definitions of close substitutes, economics has no demand functions for close substitutes to offer which can be used in a global dynamic analysis. This has obviously hampered the development of Bertrand oligopoly models for nonhomogeneous commodities.

5.2 Introduction

As we know, mathematical economics by tradition deals with local phenomena in the space of economic variables. This is due to Centuries of focus on equilibria and optima. Dynamic issues were discussed verbally, but very seldom in terms of formal models.[3]

It is difficult to extend analysis to global dynamic processes, as the traditional sketching of qualitative shapes in diagrams is of little use. One needs explicit function shapes that hold over a substantial part of the phase space, something economic theory is sparing with.

---

[2] Unfortunately, for not quite understandable reasons, Hotelling chose as his main case one where we get back to the Cournot/Bertrand original controversy. He assumed that, though the consumers would buy from the cheapest supplier (mill price accrued with transportation cost), the quantity they would buy would be independent of this total charge. As he also assumed that the competing firms would be free to choose location, they would both eventually locate in the most central location in the geographical region. As a consequence there would no longer be any difference in transportation costs, and we would be back to the original problem with the Cournot case. In a note Hotelling says that the result of locating in the same point would not hold if demand were elastic. Why Hotelling did not analyze this case is a mystery. As a matter of fact only a few years later it was proved that if the consumers but had a reservation price (buying a fixed quantity if this price was not exceeded but nothing if it was) then the paradox would evaporate.

[3] An exception were linear models of macroeconomic growth or cycles. The well-known problem with these is that they explode in terms of unbounded growth or amplitude of cycles.

A welcome exception emanates from empirical production studies in terms of Cobb-Douglas and CES functions. These were not introduced for the study of dynamic processes, rather to the purpose of organizing production data for different firms and branches in a coherent way through tuning parameters for best fit. But never mind the original purpose; given these functions work for different firms, we can also use them to study the changing circumstances over time for one or a few economic agents.

This makes them suitable for deriving demand functions in oligopoly models which have been formalized mathematically ever since their emergence with Cournot in 1838.

Cobb-Douglas and CES utility based demand functions are more suitable than linear demand, or the exponential that von Stackelberg (1934) proposed. The latter are just **assumed**, not **derived** from demand theory.

Since the success with studying nonlinear dynamic processes in the natural sciences culminated in the 1980s, oligopoly dynamics received its due share of attention.[4]

Most oligopoly models deal with the Cournot case, quantity competition in markets for homogenous commodities. In comparison quite little has been written about Bertrand oligopoly for differentiated close substitutes. See Bertrand (1883) and Edgeworth (1897).

One may wonder why this is so. The present author thinks it is due to the fact that central economic theory never produced any demand function, or even definition of substitutes that could work in a **global dynamic** setting.

5.2.1 Substitutes

Sir John Hicks (1947) presented a penetrating classification of gross/net substitutes and complements, but it all dealt with local characteristics. We do not have much help from this if we want to model a **global** dynamic process. There has simply never been presented any credible global demand function where more or less close substitutes can be modelled.[5]

[4]Notably, more engineers and mathematicians have been busy with this than economists. The danger with modellers who have better knowledge of methods of analysis than of economic facts is that formal manipulations of a few assumptions take focus to such a degree that one may wonder if they realize that economics is an empirical science.

[5]Chamberlin (1933) who wanted to replace the competitive general equilibrium theory, as the subtitle "Reconsideration of the Theory of Value" indicates, repeatedly claims that all commodities in an economy compete. Yet it is very difficult to find demand functions in the literature which actually represent this fact. Much easier to find neutral or independent demand as in the case of Cobb-Douglas type utility. In a sense all commodities included in a fixed consumer budget compete for expenditure. But technically, all expenditures are fixed budget shares, so demand only depends on the price of the commodity itself.

This is a particular disadvantage for the case of dynamic Bertrand duopoly, where one meets ad hoc assumptions about the demand function which cannot be derived from demand theory.[6]

Associated with the choice of ad hoc functions for demand, one finds a misunderstanding in modelling dynamic Bertrand duopoly.

Digression on Maximizing Profit

One may think it is important in the Bertrand case to put the profit derivative with respect to price—not quantity (thought to be associated with the Cournot case)—equal to zero.

However, whenever we have differentiated goods, we deal with monopoly markets. **Whether we find maximum profits through varying price or quantity makes absolutely no difference, as we find the same point,** provided we do things correctly.

Even in pure monopoly, as shown through Amoroso's formula, the standard is to work with inverse demand functions and optimize using the quantity variable.

5.3 A Solution: Lancaster

So, what can we do about this? There is an underutilized tool in microeconomics, i.e., Kelvin Lancaster's approach to demand theory of 1966. In essence Lancaster proposes to enter not the marketed commodities themselves, but their quantified "properties" in the utility function.

---

[6]Modellers doing nonlinear economic dynamics usually start with some ad hoc assumptions about demand functions, such as

$$\begin{aligned} q_1 &= A_1 - B_1 p_1 + C_1 p_2 \\ q_2 &= A_2 - B_2 p_2 + C_2 p_1 \end{aligned} \tag{5.1}$$

where $C_1, C_2 > 0$ indicate substitutes.

A demand function must, however, come from somewhere, be a **result of consumers maximizing utility under a budget constraint**. This is a big problem with (5.1) which **cannot fulfil any budget constraint**

$$p_1 q_1 + p_2 q_2 = y \tag{5.2}$$

Thereby we get the commodities characterized through metric properties which can be a basis for classifying more or less close substitutes.[7]

Let us consider the essentials of Lancaster's model.

5.3.1 Properties and Commodities

Denote the marketed commodities q_j and their properties x_i. Further, let a_{ij} denote the quantity of property i, that the marketed commodity j contains in each unit. Thus a desired property mix is obtained through a mix of marketed commodities:

$$x_1 = a_{11}q_1 + a_{12}q_2 \tag{5.3}$$

$$x_2 = a_{21}q_1 + a_{22}q_2 \tag{5.4}$$

It is, of course, nice to deal with full rank square matrices, $\left[a_{ij}\right]$ because they can be inverted. Lancaster by no means restricts the numbers of properties and commodities to be equal and as small as two, though for duopoly, two commodities is natural.

For the equality of rows and columns, there is a credible argument: If there are more commodities than properties, then some commodity is superfluous, i.e., dominated by a convex combination of the other ones. If, on the other hand, there are more properties than commodities, then some firm can make profit from designing a new combination of properties and market it as a new product.

Defining the determinant, assumed nonzero,

$$D = a_{11}a_{22} - a_{12}a_{21} \tag{5.5}$$

the equation system (5.3)–(5.4) can be inverted

$$q_1 = \frac{a_{22}}{D}x_1 - \frac{a_{12}}{D}x_2 \tag{5.6}$$

$$q_2 = \frac{a_{11}}{D}x_2 - \frac{a_{21}}{D}x_1 \tag{5.7}$$

[7]In this format we can also incorporate a product design problem if we let the design in terms of properties be an additional means of competition. To this end we need to develop another piece of microeconomics, i.e., the theory of joint production.

Lancaster suggested that the utility function to be maximized should have x_i, the properties, as arguments, not q_j, the commodities.[8]

5.3.2 Utility

For utility it is convenient to take a usual Cobb-Douglas function $U = x_1^{\alpha_1} x_2^{\alpha_2}$. As there is enough complexity in the model, we skip asymmetries and take $\alpha_1 = \alpha_2$, so[9]

$$U = \sqrt{x_1 x_2} \tag{5.8}$$

The arguments x_1, x_2 are precisely the properties we spoke of.

5.3.3 Budget Constraint

The consumers maximize utility (5.8) subject to system (5.3)–(5.4) and a budget constraint:

$$p_1 q_1 + p_2 q_2 = 1 \tag{5.9}$$

To avoid unnecessary symbols, income is normalized to unity.[10]

The budget constraint is in terms of commodity quantities q_1, q_2, and their prices, **not the properties** x_1, x_2, which were entered in the utility function (5.8).

Commodity prices are denoted p_1, p_2—for the properties we do not, as yet, have any prices.

---

[8]This also opens a wider perspective—as we get shadow prices for the properties, we can also extend models to choose design for the competing products. Hopefully, in a global dynamics, we could then not only find out the orbits of changing commodity prices, but also study the design of the competing products. It would be interesting to know if the competitors would asymptotically converge upon the same design, or, rather choose very different designs. Other attractors than fixed points, for instance periodic, would be even more interesting, as one could then model fashion cycles.

[9]With ordinal utility we could skip the square root sign, but it does not make any harm either.

[10]Note that an advantage of the Cobb-Douglas function is that we can have any number of different consumers and yet construct a utility for the entire market as if we dealt with one single consumer.

5.3.4 *Optimization*

We now have to maximize (5.8) subject to (5.3)–(5.4) and (5.9). To this end, associate Lagrange multipliers λ_1, λ_2 with Eqs. (5.3)–(5.4). Not unexpectedly they turn out to be shadow prices for the properties x_1, x_2.[11]

Optimizing with respect to x_1, x_2 we get

$$\frac{1}{2}\sqrt{\frac{x_2}{x_1}} = \lambda_1 \tag{5.10}$$

$$\frac{1}{2}\sqrt{\frac{x_1}{x_2}} = \lambda_2 \tag{5.11}$$

and with respect to q_1, q_2

$$a_{11}\lambda_1 + a_{21}\lambda_2 = p_1 \tag{5.12}$$

$$a_{12}\lambda_1 + a_{22}\lambda_2 = p_2 \tag{5.13}$$

Multiplying Eqs. (5.12)–(5.13) by q_1, q_2 respectively, adding, and using (5.3)–(5.4) we get

$$\lambda_1 x_1 + \lambda_2 x_2 = p_1 q_1 + p_2 q_2 = 1 \tag{5.14}$$

Both (5.12)–(5.13) and (5.14) demonstrate that λ_1, λ_2 are **imputed property prices**.

It is convenient to solve (5.12)–(5.13) for λ_1, λ_2

$$\lambda_1 = \frac{a_{22}}{D} p_1 - \frac{a_{21}}{D} p_2 \tag{5.15}$$

$$\lambda_2 = \frac{a_{11}}{D} p_2 - \frac{a_{12}}{D} p_1 \tag{5.16}$$

In optimum, $U = \sqrt{x_1 x_2} = 1$, so (5.10)–(5.11) simplify to

$$x_1 = \frac{1}{2\lambda_1} \tag{5.17}$$

$$x_2 = \frac{1}{2\lambda_2} \tag{5.18}$$

[11]The budget constraint also needs a Lagrange multiplier, but it has no significance for the model and can be absorbed as a factor in λ_1, λ_2.

5.3.5 Demand Functions

We can now substitute λ_1, λ_2 from (5.15)–(5.16) in (5.17)–(5.18), and then the so obtained x_1, x_2 in (5.6)–(5.7). So finally

$$q_1 = \frac{1}{2p_1}\left(\frac{1}{1-\frac{a_{21}}{a_{22}}\frac{p_2}{p_1}} + \frac{1}{1-\frac{a_{11}}{a_{12}}\frac{p_2}{p_1}}\right) \tag{5.19}$$

$$q_2 = \frac{1}{2p_2}\left(\frac{1}{1-\frac{a_{22}}{a_{21}}\frac{p_1}{p_2}} + \frac{1}{1-\frac{a_{12}}{a_{11}}\frac{p_1}{p_2}}\right) \tag{5.20}$$

Note that if $q_1 < 0$ then we must set $q_1 = 0$ and $q_2 = \frac{1}{p_2}$. Likewise, if $q_2 < 0$ then $q_2 = 0$ and $q_1 = \frac{1}{p_1}$. In either case only one commodity will be demanded, i.e., we get a monopoly. This is a problematic case for isoelastic demand.[12]

Even in duopoly this case is problematic. The reaction functions have infinite slope at their intersection in the origin, so this fixed point is as unstable as anything can be. Yet, whenever the orbit happens to hit the origin it sticks there. See Tramontana et al. (2010). Any constraint or random perturbation would immediately repel the orbit from the origin, but modelling becomes problematic because the process crucially depends on the exact type of repulsion mechanism assumed. Therefore it is generally wise to assume adaptive behaviour in these types of models to prevent them from ever hitting the origin.[13]

5.3.6 Inverse Demand

In order to write the demand functions in a more succinct form define new coefficients

$$b_1 = \frac{a_{11}}{a_{12}} \tag{5.21}$$

$$b_2 = \frac{a_{22}}{a_{21}} \tag{5.22}$$

[12]As a monopolist then gets revenue $p_j q_j = 1$ and incurs cost $c_j q_j$, its best choice is to produce nothing, sell the produce at an infinite price, and avoid any production cost.

[13]This is normally done by assuming that the competitors only move part of the way from the previous move to the new calculated best reply.

They represent, for each property, the ratios to which the two products of marketed goods contain these in their design. Note that b_1, b_2 depend on design decisions by **both** competitors, and also that they are defined in an asymmetric way.

Using (5.21)–(5.22), (5.19)–(5.20) can be written[14]:

$$q_1 = \frac{1}{2}\frac{b_2}{b_2 p_1 - p_2} + \frac{1}{2}\frac{1}{p_1 - b_1 p_2} \tag{5.23}$$

$$q_2 = \frac{1}{2}\frac{1}{p_2 - b_2 p_1} + \frac{1}{2}\frac{b_1}{b_1 p_2 - p_1} \tag{5.24}$$

As indicated, profit maximum in monopolistic competition is better formulated in the style of Amoroso, so we want to use the inverse demand functions

$$p_1 = \frac{1}{2}\frac{b_1}{b_1 q_1 + q_2} + \frac{1}{2}\frac{1}{q_1 + b_2 q_2} \tag{5.25}$$

$$p_2 = \frac{1}{2}\frac{1}{b_1 q_1 + q_2} + \frac{1}{2}\frac{b_2}{q_1 + b_2 q_2} \tag{5.26}$$

Cournot Case

Note that (5.25)–(5.26) embeds the case of Cournot duopoly, provided the commodities are identical. To see this, substitute $b_1 = b_2 = 1$ to obtain

$$p_1 = p_2 = \frac{1}{q_1 + q_2} \tag{5.27}$$

Hence with this particular design combination the model is capable of all dynamics that isoelastic Cournot duopoly is capable of. However, the use of Lancaster's model breaks down in this case, as the system determinant (5.5) is degenerate, so none of the inversions discussed in the beginning is possible. This case has been extensively discussed in the present author's work on Cournot, Puu (1991).

[14]Of course, only restricted price ratios provide for positive demand of both commodities. Most succinctly the condition can be expressed in only two compound variables, $b_1 \frac{p_2}{p_1}$ and $b_2 \frac{p_1}{p_2}$,

$$-2 < \frac{\left(b_1 \frac{p_2}{p_1} - 1\right) + \left(\frac{1}{b_2}\frac{p_2}{p_1} - 1\right)}{(b_1 \frac{p_2}{p_1} - 1)\cdot(\frac{1}{b_2}\frac{p_2}{p_1} - 1)} < 0$$

In this interval the commodities are substitutes. If any of Eqs. (5.19)–(5.20) is negative, then only one commodity is demanded, so $q_1 = \frac{1}{p_1}, q_2 = 0$, or $q_1 = 0, q_2 = \frac{1}{p_2}$, the awkward case commented above.

5.4 Duopoly

Of course we assume that in duopoly different firms produce the competing products. With the demand functions given, we can set up profit maximization, and, hopefully, derive reaction functions also for the Bertrand cases with distinct but substitutive commodities.

Assume fixed commodity designs in terms of the $\left[a_{ij}\right]$ in order to study a pure price/quantity dynamics. To this end also assume a production function, again linearly homogenous, symmetric Cobb-Douglas,

$$q_j = \sqrt{k_j l_j} \tag{5.28}$$

Suppose we take q_j as given and minimize production cost

$$C_j\left(q_j\right) = r_j k_j + w_j l_j \tag{5.29}$$

subject to (5.28). The solution is textbook standard:

$$k_j = \frac{c_j}{r_j} q_j \tag{5.30}$$

$$l_j = \frac{c_j}{w_j} q_j \tag{5.31}$$

where

$$c_j = 2\sqrt{r_j w_j} \tag{5.32}$$

Hence

$$C_j\left(q_j\right) = c_j q_j \tag{5.33}$$

Note that we took input prices r_j, w_j *different* for the oligopolists in order to make them differ in terms of marginal cost. Otherwise, given the firms access the same technology, they become identical, synchronize, and produce but trivial results.

5.4.1 Amoroso Formulas

Profits are now $p_j\left(q_1, q_2\right) q_j - C_j\left(q_j\right)$ where inverse demand $p_j\left(q_1, q_2\right)$ was derived in (5.25)–(5.26) above. Using Amoroso's formula, the competitors maximize profits setting

$$p_j + q_j \frac{\partial p_j}{\partial q_j} = \frac{dC_j}{dq_j} \tag{5.34}$$

Using (5.25)–(5.26) and (5.33)

$$\frac{1}{2}\frac{b_1 q_2}{(b_1 q_1 + q_2)^2} + \frac{1}{2}\frac{b_2 q_2}{(q_1 + b_2 q_2)^2} = c_1 \tag{5.35}$$

$$\frac{1}{2}\frac{b_1 q_1}{(b_1 q_1 + q_2)^2} + \frac{1}{2}\frac{b_2 q_1}{(q_1 + b_2 q_2)^2} = c_2 \tag{5.36}$$

The first equation serves to determine $q_1 = \phi_1(q_2)$, the second $q_2 = \phi_2(q_1)$. That the denominators are equal is deceptive, because the optimum choice of q_1 in the first and q_2 in the second, will make them different in most situations.[15]

5.5 Reaction Functions

As indicated, we can **not** solve for the reaction functions. However, the two terms in either equation in (5.35)–(5.36) produce very similar results, similar to those that appear in the Cournot cases, so we could solve for the reaction function either using the first or the second term. Doing this, we get

$$q_i = \sqrt{A_i}\sqrt{\frac{q_j}{2c_i}} - A_i q_j \tag{5.37}$$

where

$$A_1 = \frac{1}{b_1} \vee b_2$$

$$A_2 = \frac{1}{b_2} \vee b_1$$

As we see, choosing either of the two terms, we get the same unimodal shape of reaction function, starting from the origin with a leading square root term, reaching a maximum, and then getting down to zero due to the eventually dominant linear term.

Numerical algorithms to determine best replies result in precisely such unimodal shapes as produced by a leading square root term, eventually dominated by a linear term.[16] Figure 5.1 illustrates what the reaction functions look like for various design parameters.

[15]We may, however, also take Eqs. (5.35)–(5.36) as a simultaneous system, which produces a fixed point. To illustrate assume identical goods, i.e., $b_1 = b_2 = 1$. Then $q_1 = \frac{c_2}{(c_1+c_2)^2}$, $q_2 = \frac{c_1}{(c_1+c_2)^2}$, which are well known coordinates of the Cournot point.

[16]Quite as in the logistic process, though there we deal with linear and quadratic terms.

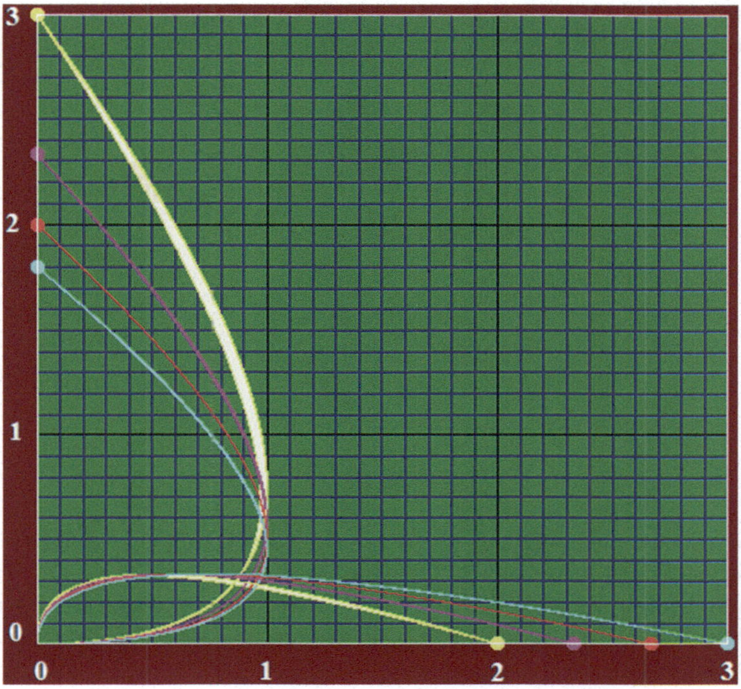

Fig. 5.1 Best reply reaction functions for the competitors. Parameters: $c_1 = 0.25$, $c_2 = 0.75$, $b_1 = 0.5$; $b_2 = 1.0$ (yellow), 1.5 (violet), 2.0 (orange), 2.5 (blue). The four cases generated by varying b_2 belong pair-wise together. Notably, whereas one function is stretched out, the other is contracted. It is also noteworthy that the maxima of the functions are about the same height in each family, though reciprocal to marginal cost—for the first firm the height is about 1 for the second whose marginal cost is thrice as high it is about 1/3. Further note that the curves are generated twice, once by a profit maximizing search algorithm, once by an approximation described below. Only careful scrutiny shows that some of the doubly generated curves are a bit "thick", and only the yellow curves show up "separated threads". Visually the approximations are rather accurate

They are calculated by a numerical algorithm to maximize profits according to (5.25)–(5.26) and (5.33). To generate these specimens we kept $c_1 = 0.25$, $c_2 = 0.75$, and $b_1 = 0.5$ constant, and changed b_2 in the range of 1.0–2.5. In steps of 0.5. Note that the changing design coefficient shrinks one reaction function **down**, while stretching the other to the **right**. The colour code also shows how the reaction functions are coupled together.

As we will see, there is truly detailed agreement between numerical results and the particular square root/linear combination. (1) The maximum hump is at **one fourth** of the total positive stretch of the reaction functions. (2) The height of this hump is **reciprocal to marginal cost** c_i for the supplier and **does not depend on any other parameter**.

This gives us a good prospect for approximating the reaction functions we cannot solve for from (5.35)–(5.36). The present model was first proposed in Puu (2017).

5.5.1 *Approximating the Reaction Functions*

Note that we can obtain a few pieces of information from (5.35)–(5.36):

(i) By substituting zeros for q_1, q_2 in the respective equations, we can solve for the remaining variable:

$$q_2 = \frac{1 + b_1 b_2}{2b_2} \frac{1}{c_1} \tag{5.38}$$

$$q_1 = \frac{1 + b_1 b_2}{2b_1} \frac{1}{c_2} \tag{5.39}$$

These mark the endpoints of the positive stretch of the reaction functions and are indicated by the small dots in Fig. 5.1. Note that they are calculated **analytically** from (5.35)–(5.36), whereas the function graphs were obtained **numerically**. Yet they fit exactly to the numerically obtained endpoints. What else do we have?

(ii) The locations of maxima at **one fourth** distance of the positive stretch:

$$q_2 = \frac{1 + b_1 b_2}{8b_2} \frac{1}{c_1} \tag{5.40}$$

$$q_1 = \frac{1 + b_1 b_2}{8b_1} \frac{1}{c_2} \tag{5.41}$$

So, let us substitute these in (5.35)–(5.36) in order to find the heights of the maximum humps.

$$\frac{b_1 b_2}{(b_1 b_2 + 8b_1 b_2 c_1 q_1 + 1)^2} + \frac{1}{(b_1 b_2 + 8c_1 q_1 + 1)^2} = \frac{1}{4} \frac{1}{1 + b_1 b_2} \tag{5.42}$$

$$\frac{1}{(b_1 b_2 + 8c_1 q_1 + 1)^2} + \frac{b_1 b_2}{(b_1 b_2 + 8b_1 b_2 c_1 q_1 + 1)^2} = \frac{1}{4} \frac{1}{1 + b_1 b_2} \tag{5.43}$$

after some reorganization.

Solving for q_1 and q_2 respectively, we get estimates of the heights of the maximum humps of the reaction functions. This time, unlike the case of (5.40)–(5.41), we do not obtain any handy expressions. If we try the most potent CAS, we arrive at page-long compositions of roots to high degree polynomials.

However, there are two interesting facts about the solutions (5.42)–(5.43): The two equations are **identical**, so $c_1 q_1 = c_2 q_2$ (the lower the marginal cost the higher the maximum hump), quite as we found numerically, and, the equations do not contain b_1 and b_2 separately, **only in the composition** $b_1 b_2$, so whenever we fix this product, we get the same value for $c_1 q_1 = c_2 q_2$.

These are hence good reasons for taking an approximation for reaction functions in terms of the square root/linear combination.

On the Fine Tuning

In what follows, we show bifurcation diagrams for $b_1 b_2 = 0.8$. For numerical values, we **can** solve (5.42)–(5.43), so for 0.8 we get $c_1 q_1 = c_2 q_2 \approx 0.2492$.

Defining amplitudes

$$A_1 = \frac{2b_2}{1 + b_1 b_2} \tag{5.44}$$

$$A_2 = \frac{2b_1}{1 + b_1 b_2} \tag{5.45}$$

we take the approximation

$$q_1 = \sqrt{A_1}\sqrt{\frac{q_2}{c_1}} - A_1 q_2 \tag{5.46}$$

$$q_2 = \sqrt{A_2}\sqrt{\frac{q_1}{c_2}} - A_2 q_1 \tag{5.47}$$

If we take other values for $b_1 b_2 = \frac{A_1 A_2}{4 - A_1 A_2}$, we may have to choose other amplitudes in the (5.46)–(5.47) formulas, though moderate variation of $b_1 b_2 \in [0.5, 2.0]$ causes very little change of $c_1 q_1 = c_2 q_2 \in [0.2422, 0.2500]$.[17] Therefore the numerical approximation is **rather robust** even if we keep to (5.46)–(5.47) without further fine tuning.

The general shape of the reaction functions remains the same. We added the **approximations** as well to Fig. 5.1. Compare the doubled graphs displayed. In most cases we only see the twice generated curves as a bit "thick", in just one case the curves seem separated in threads.

[17] The way we set up the b_1, b_2 coefficients implies that a product deviating very much from unity would require the design to be extremely deficient or extremely overpowered in terms of secondary properties, so this is hardly a normal situation.

So we can confidently proceed analysis with the approximate reaction functions.[18]

Now, having (5.46)–(5.47), we could go ahead and try to calculate the fixed point coefficients, differentiate, obtain Eigenvalues, and note the loss of stability through a Neimark-Sacker bifurcation.

5.5.2 Fixed Point Destabilization

Given the shape of the reaction functions (5.46)–(5.47), there is just one off origin fixed point. If it were a case of Cournot oligopoly we would call it the Cournot point; now we have no specific name for it. The destabilization of this unique equilibrium is of particular interest.

As is the case with so many algebraic problems in this model, we cannot solve for the fixed point coordinates in general from (5.46)–(5.47).

But we can put (5.46)–(5.47) in the format

$$\left(\frac{q_1}{q_2} + A_1\right)^2 = \frac{A_1}{c_1 q_2} \tag{5.48}$$

$$\left(\frac{q_2}{q_1} + A_2\right)^2 = \frac{A_2}{c_2 q_1} \tag{5.49}$$

In principle (5.48)–(5.49) could serve to determine q_1, q_2, but unfortunately the resulting polynomial equations are too high degree to provide any informative results. However, (5.48)–(5.49) are useful in one way: Once we know the ratios $\frac{q_1}{q_2}, \frac{q_2}{q_1}$, we can easily calculate the numerator and the denominator q_1, q_2.

Now consider the fixed points just at the **threshold of destabilization**.

Differentiating (5.46)–(5.47) we get the derivatives

$$\frac{\partial q_1}{\partial q_2} = \frac{1}{2}\sqrt{A_1}\sqrt{\frac{1}{c_1 q_2}} - A_1 \tag{5.50}$$

$$\frac{\partial q_2}{\partial q_1} = \frac{1}{2}\sqrt{A_2}\sqrt{\frac{1}{c_2 q_1}} - A_2 \tag{5.51}$$

which are also the Eigenvalues.

[18] A consolation for having to work with approximations is that the competitors on the market we want to portray hardly can do any better, as no available CAS (computer algebra system) can solve (5.35)–(5.36).

Using (5.46)–(5.47) anew, (5.50)–(5.51) can be written

$$\frac{\partial q_1}{\partial q_2} = \frac{1}{2}\left(\frac{q_1}{q_2} - A_1\right) \tag{5.52}$$

$$\frac{\partial q_2}{\partial q_1} = \frac{1}{2}\left(\frac{q_2}{q_1} - A_2\right) \tag{5.53}$$

At fixed point destabilization we have[19]

$$\frac{\partial q_1}{\partial q_2}\frac{\partial q_2}{\partial q_1} = -1 \tag{5.54}$$

which using (5.52)–(5.53) can be written

$$A_1\frac{q_2}{q_1} + A_2\frac{q_1}{q_2} = 5 + A_1 A_2 \tag{5.55}$$

This is only second order in the ratio $\frac{q_1}{q_2}$, or $\frac{q_2}{q_1}$, and can be easily solved for, say,

$$\frac{q_2}{q_1} = \frac{1}{2A_1}\left(5 + A_1 A_2 \pm \sqrt{25 + 6A_1 A_2 + A_1^2 A_2^2}\right) \tag{5.56}$$

As noted, substituting in (5.48)–(5.49), we can finally solve for the coordinates q_1 and q_2 individually:

$$q_1 = 4\frac{A_1}{c_2}\frac{A_1 A_2}{5 + 3A_1 A_2 \pm \sqrt{25 + 6A_1 A_2 + A_1^2 A_2^2}} \tag{5.57}$$

$$q_2 = 4\frac{A_2}{c_1}\frac{A_1 A_2}{5 + 3A_1 A_2 \pm \sqrt{25 + 6A_1 A_2 + A_1^2 A_2^2}} \tag{5.58}$$

As the denominators are positive **for both roots**, both can provide solutions. Which one to choose depends on which renders the highest profit. Provided we take the positive or negative signs in **both** Eqs. (5.57)–(5.58), we get from (5.52)–(5.53) the handy expressions

$$\frac{\partial q_1}{\partial q_2} = \frac{1}{2}\frac{A_1}{A_2}\left(\frac{c_1}{c_2} - A_2\right) \tag{5.59}$$

$$\frac{\partial q_2}{\partial q_1} = \frac{1}{2}\frac{A_2}{A_1}\left(\frac{c_2}{c_1} - A_1\right) \tag{5.60}$$

[19]From the approximate reaction functions we might expect a flip bifurcation, however, the adaptive format transforms the case to a Neimark-Sacker.

If, we take opposite signs in (5.57)–(5.58), we, on the contrary, get extremely nasty algebra, which can be used, though is unsuitable for reproduction in print.

Considering again (5.54), we find the bifurcation boundary in parameter space

$$A_1 \frac{c_1}{c_2} + A_2 \frac{c_2}{c_1} = 5 + A_1 A_2 \tag{5.61}$$

or, using (5.44)–(5.45),

$$2b_2 \frac{c_1}{c_2} + 2b_1 \frac{c_2}{c_1} = 5 + 4\frac{b_1 b_2}{1 + b_1 b_2} \tag{5.62}$$

which we will utilize to calculate a bifurcation boundary in the numerics for a bifurcation diagram in b_1, c_2-space. Not surprisingly it becomes a hyperbola, though recall that for other bifurcation diagrams we might need different branches of (5.57)–(5.58) for which (5.62) does not serve.

Let us now take a look at a numerical study.

5.6 Numerical Results

To avoid the conceptually problematic monopoly cases we choose the adaptive formulation, though with very high weights, 99.5%, for the calculated best reply and only 0.5% for the previous move.

5.6.1 Computation Model

Setting up the formal iterative model is now easy:

$$q_1' = (1 - \delta)\, q_1 + \delta \left(\sqrt{\frac{2b_2}{1 + b_1 b_2}} \sqrt{\frac{q_2}{c_1}} - \frac{2b_2}{1 + b_1 b_{22}} q_2 \right) \tag{5.63}$$

$$q_2' = (1 - \delta)\, q_2 + \delta \left(\sqrt{\frac{2b_1}{1 + b_1 b_2}} \sqrt{\frac{q_2}{c_2}} - \frac{2b_1}{1 + b_1 b_2} q_1 \right) \tag{5.64}$$

As usual the dash indicates a one period forward iteration.

We take $\delta = 0.995$ fixed throughout the numerical study.[20] There remain four parameters, b_1, b_2, c_1, c_2, representing design coefficients and marginal costs. From

[20] A more realistic assumption would be to use different adaptation coefficients δ_1, δ_2 for the competitors, but we have more parameters than we can visualize anyhow, so for a first study it is always better to focus on those that are most essential.

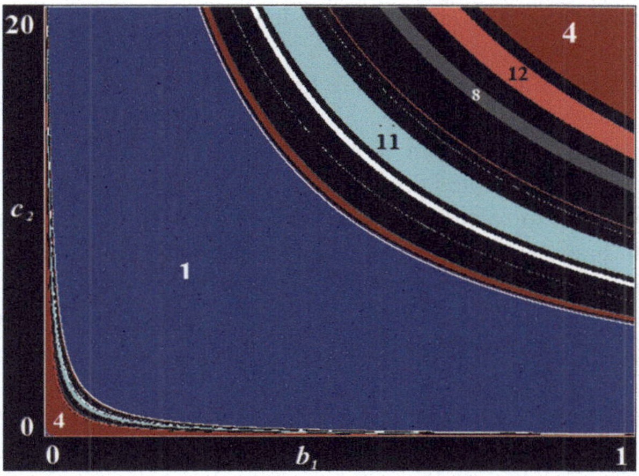

Fig. 5.2 Arnol'd tongues for the 15 lowest periodic solutions to the approximating system (5.63)–(5.64) with parameters $c_1 = 1$ and $1 < c_2 < 20$, $0.05 < b_1 < 1$ (given $b_1 b_2 = 0.8$). The asymptotic orbits were checked for the 15 lowest periodicities. Dominant, except the fixed point is period 4

models of similar structure of Cournot duopoly, we know that the c_1, c_2 only influence the outcome through their ratio $\frac{c_2}{c_1}$. We can therefore also fix $c_1 = 1$, which leaves us with three independent parameters, still too many for complete bifurcation diagrams.

Figure 5.1 shows the bifurcation diagram for a combined variation of the cost ratio and design parameters. Fix $b_1 b_2 = 0.8$ and $c_1 = 1$. Then choose ranges $0 < b_1 < 1$ (horizontal) and $0 < c_2 < 20$ (vertical), and check for the 15 lowest periodicities.

The colour code, after the fixed point (dark blue) shows periodicities 4 (brown), 15 (white), 11 (turquoise), 8 (gray), 12 (red), and 4 (brown) again.[21]

In Fig. 5.2 we also inserted parabola shaped curves to indicate destabilization of the fixed point. These are calculated as follows: Given $b_1 b_2 = 0.8$ and $c_1 = 1$ we from definitions have $A_1 = \frac{8}{9}\frac{1}{b_1}$ and $A_2 = \frac{10}{9} b_1$. Substituted in (5.62) we find two solutions for the parameters in our bifurcation plane

$$b_1 c_2 = \frac{97}{36} \pm \frac{\sqrt{209305}}{180}$$

[21] Note that a reversed and compressed version of the upper right picture appears in the lower left corner. This is because, as noted, only the ratio of marginal costs matters. Having fixed $c_1 = 1$, c_2 and $\frac{1}{c_2}$ hence influence the system in the same way on either side of unity.

Note how well these two analytically calculated Neimark-Sacker parabolas fit the destabilization boundaries in Fig. 5.2.

There remain black bands containing higher periods, quasiperiodicity, or even chaos.

More irregularity can be seen if we draw the bifurcation diagram in design parameter space b_1, b_2, keeping costs constant as in Fig. 5.3. the structure becomes folded, so if we want to calculate the bifurcation boundary in such cases, we have to use different branches of (5.57)–(5.58), which is computationally complicated, though a good exercise. We also indicated areas of positive highest Lyapunov exponent (or rather > 0.05 to be quite certain) in the same diagram, in the hatched golden shade. Note how well it fits between the coloured period tongues of the picture. The high and irregular periodicities are also noteworthy.

We may also note a certain overlap of periodic and (possibly) chaotic belts. This is no anomaly. In similar models there often is coexistence of attractors, for instance periodic and chaotic. Even the eventual destabilization of the fixed point is by a subcritical Neimark-Sacker bifurcation where the loss of stability is due to a vanishing basin of attraction. Several studies on such phenomena have been produced. See Anna Agliari et al. (2005).

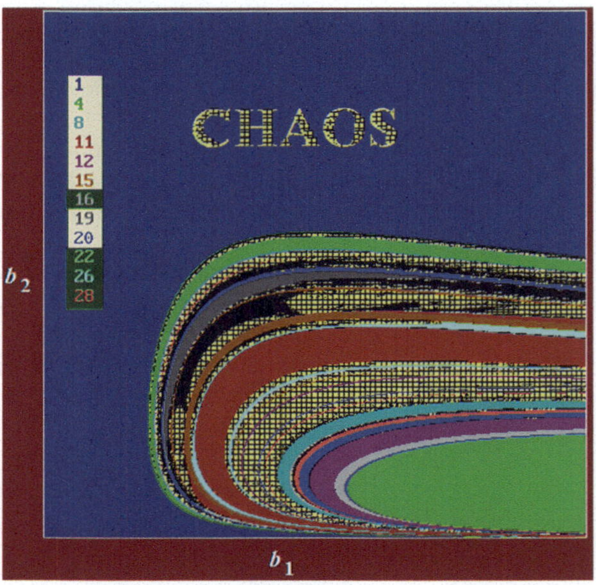

Fig. 5.3 The bifurcation diagram in design parameter space b_1, b_2. Noteworthy is the folded period structure and the predominance of chaos in the system. This may indicate that we might indeed find some interesting dynamics once we make product design endogenous in the model

5.7 Product Design

5.7.1 Introduction

It is now time to include the product design issue. As we can calculate the imputed property prices at each stage of the dynamic orbit, we also get the possibility to evaluate each design.

It may thus occur that a change in design of any competing product may be profitable. Along with price, the firms might then also use product design as a means of competition.

To accomplish this we only need to develop the theory of joint production along the lines suggested by Frisch (1965) (available in Norwegian since 1927, and plagiarized by several authors in German and English who had the advantage of knowing the language).[22]

Once the design problem is formalized, there are several highly interesting issues to study: Do the competitors eventually converge upon a design of identical products? Or, do they approach different designs? The case of periodic attractors is even more interesting: Can we explain recurrent fashion cycles in the design of products?

We need to develop the theory for joint production a little. The theory of joint production was developed by the first economics Nobel Laureate Ragnar Frisch. His is the best taxonomy of cases we ever got in economics. However, the format is graphical, and we do not find any useful form of function for joint products that we can use in global analysis. We can just lean on the idea that what we need is a case of "assortment" of degree one (number of products minus number of functional constraints).

Once we consider substitution between joint products, we can also think of the product characteristics as possible Lancasterian properties on the productions side. Thinking in terms of putty/clay modelling of the 1970s the property variables x_1, x_2 can be chosen on a continuous production possibility frontier (putty stage), but be frozen in fixed proportions once a design choice is made (clay stage). Again, the Norwegian school provides the best ideas in the work by Leif Johansen.

[22]The combination of properties bundled together in a marketed product can easily be seen as parallel to joint production. The transition from production possibilities to a frozen mix of properties can then itself be modelled through "putty/clay" ideas from the 1970s—best presented by Johansen in 1972, another important achievement of the Oslo school.

5.8 The Model

5.8.1 The Production Possibility Frontier

A convenient joint production function, CES with $\rho = \frac{1}{2}$ could read

$$\sqrt{x_{1j}^2 + x_{2j}^2} = q_j \tag{5.65}$$

where x_{1j}, x_{2j} are the outputs for joint production, and q_j depends on the inputs of capital and labour $q_j = \sqrt{k_j l_j}$, as suggested in (5.28). Despite notation, q_j at present only denote convenient intermediate variables as long as substitution along the possibility frontier (5.65) is open. It is the x_{ij} that are the products. Though, once a mix of joint products is frozen, the q_j become products with imputed properties $a_{ij} = x_{ij}/q_j$.

Design Coefficients

Once the competitors decide a design, we impute the properties to the product unit, i.e., set

$$x_{1j} = a_{1j} q_j \tag{5.66}$$

$$x_{2j} = a_{2j} q_j \tag{5.67}$$

5.8.2 Optimum Design

To choose a property mix, we will now have good use of the imputed property prices λ_1, λ_2. The "revenue" is then $\lambda_1 x_{1j} + \lambda_2 x_{2j}$ and the cost, as usual, $r_j k_j + w_j l_j$. Recall that we assumed different capital and labour costs for the competitors, the reason being that we need some difference between them so as to avoid trivial outcomes when the firms can access the same technology.

If we want to maximize the difference $\lambda_1 x_{1j} + \lambda_2 x_{2j} - r_j k_j - w_j l_j$ subject to the two components for the production possibility frontier (5.65) and (5.28), we will not get any definite optimum for x_{1j}, x_{2j} because the maximand is linear and the combined constraints are linear homogenous.

However, substituting from (5.66)–(5.67) for the "joint products" in (5.65) we get $\sqrt{a_{1j}^2 q_j^2 + a_{2j}^2 q_j^2} = q_j$, or cancelling q_j,

$$\sqrt{a_{1j}^2 + a_{2j}^2} = 1 \tag{5.68}$$

Using this constraint (5.68) we can maximize the value

$$\lambda_1 a_{1j} + \lambda_2 a_{2j} \tag{5.69}$$

of the unit product in terms of design.[23]

This design optimization has the very simple solution

$$a_{1j} = \frac{\lambda_1}{\sqrt{\lambda_1^2 + \lambda_2^2}} \tag{5.70}$$

$$a_{2j} = \frac{\lambda_2}{\sqrt{\lambda_1^2 + \lambda_2^2}} \tag{5.71}$$

Obviously the optimal design at each step of the orbit is **the same** for **both** competitors, as it should, given they share the same technology and face the same property values.

Whether this means that the competitors converge on identical designs in the end, remains to be seen. Most likely the competitors will not change design every period. We need an assumption about when they do and how much. One possibility is to assume that they do when they find that actual design in terms of value falls too short of what is optimal. Hence they would not adjust simultaneously, and if the attracting orbit is more complex, they might get out of phase and settle on different orbits for the design.

A Useful Property

Consider a particular property of the optimal design coefficients. According to (5.70)–(5.71) **they lie on a unit circle**, and can hence be represented by **a single angular coordinate** θ_j. Hence we can write

$$a_{1j} = \cos\theta_j, \; a_{2j} = \sin\theta_j \tag{5.72}$$

This is not so exciting in itself, but, as the competitors update their designs to the optimal **at some stage** of the process, the **actual design** after this first update will always **remain on the unit circle**. The design **updates can hence be represented through the single variable** θ_j.

[23] The optimal mix of inputs is a problem we already solved (5.30)–(5.32).

Considering the composite design parameters used above we get from (5.21)–(5.22)

$$b_1 = \frac{\cos\theta_1}{\cos\theta_2} \tag{5.73}$$

$$b_2 = \frac{\sin\theta_2}{\sin\theta_1} \tag{5.74}$$

Hence from (5.44)–(5.45)

$$A_1 = \frac{\sin(2\theta_2)}{\sin(\theta_1 + \theta_2)} \tag{5.75}$$

$$A_2 = \frac{\sin(2\theta_1)}{\sin(\theta_1 + \theta_2)} \tag{5.76}$$

using some fundamental identities in trigonometry. Fortunately, whenever $\cos\theta_j$ and $\sin\theta_j$ are positive the amplitudes A_1 and A_2 as well are positive, which is good to know as one never knows with trigonometry.

Further, we can obtain the imputed property values λ_1, λ_2 using again the convenient (5.72). Even if the firms do not update their design every period, they need to make an estimate of whether it is worthwhile. Therefore they always need the imputed property values λ_1, λ_2.

Using (5.25)–(5.26) and (5.73)–(5.74) in (5.15)–(5.16) we get

$$\lambda_1 = \frac{1}{2}\frac{1}{q_1\cos\theta_1 + q_2\cos\theta_2} \tag{5.77}$$

$$\lambda_2 = \frac{1}{2}\frac{1}{q_1\sin\theta_1 + q_2\sin\theta_2} \tag{5.78}$$

These are the imputed prices expressed as dependent on the supplies of the competitors and their angular design variables.

How they depend on commodity prices on the market was stated above in (5.12)–(5.13). Further, our iterative map (5.65)–(5.66) was formulated for **quantities and not prices**.

It was never necessary to calculate the prices, which, however is easily done using (5.27)–(5.28) along with (5.73)–(5.74) for b_1, b_2—originally fixed parameters along with c_1, c_2—but now endogenous variables.

The derivations, however, are quite messy, so, in order not to break the thread of reasoning, we skip the details, only writing the most important results.[24]

Dividing (5.78) by (5.77) we then, using (5.70)–(5.71) and (5.72), immediately get the optimal design angle

$$\theta = \arctan \frac{\lambda_2}{\lambda_1} \tag{5.79}$$

in case any competitor goes for design change. We are lucky as $0 < \frac{\lambda_2}{\lambda_1} < \infty$ according to (5.79) returns $0 < \theta < \frac{\pi}{2}$.

Once the process has settled on the unit circle, from (5.72), we can calculate the current value of the design:

$$V_j = \lambda_1 \cos \theta_j + \lambda_2 \sin \theta_j \tag{5.80}$$

and compare it to the optimal value

$$U = \lambda_1 \cos \theta + \lambda_2 \sin \theta \tag{5.81}$$

where θ was stated in (5.79).

Of course, all this convenience depends on assuming a CES function with $\rho = \frac{1}{2}$, which may seem like a trick. However, what is modelling if not using such tricks to simplify the analysis of complex problems. The production possibility frontier as a circular arc enables the use of trigonometry. The method does not work for $\rho = 0.4999$ nor for $\rho = 0.5001$, but as the circular arc represents the qualitative features we want, why not use it?

5.8.3 Design Update Rule

For the update decision define a coefficient $k > 1$, such that the competitor changes design whenever

$$U > kV_j \tag{5.82}$$

[24] With this kind of derivations it is good to use some kind of computer algebra, so called CAS—there are quite a number, both free on the web, such as Maxima, and very costly, such as Maple and Mathematica. These have two advantages: (1) They do no trivial mistakes such as one always does. (2) They have all the standard mathematical formulas in their memory. However, they are not intelligent. One must always use one's own intuition for the solution to help them along. Leaving them to simplify or solve something always results in some uselessly messy expression or leaves the computer labouring for ever.

setting $\theta'_j = \arctan\frac{\lambda_2}{\lambda_1}$, otherwise retaining $\theta'_j = \theta_j$. If, for instance $k = 1.5$ then design is changed whenever the potential value of the change overshoots the actual by 50%. Keeping k low increases the probability that the competitors converge to the same design. Then we are back at the original model, though with the difference that product design is now an outcome of the dynamic process.[25]

5.9 The Design Update Map

We now have all the bits and pieces needed for formulating the iterative map including both price and design competition.[26]

$$q_1' = (1-\delta)\,q_1 + \delta\left(\sqrt{\frac{\sin(2\theta_2)}{\sin(\theta_1+\theta_2)}}\sqrt{\frac{q_2}{c_1}} - \frac{\sin(2\theta_2)}{\sin(\theta_1+\theta_2)}q_2\right)$$

$$q_2' = (1-\delta)\,q_2 + \delta\left(\sqrt{\frac{\sin(2\theta_1)}{\sin(\theta_1+\theta_2)}}\sqrt{\frac{q_2}{c_2}} - \frac{\sin(2\theta_1)}{\sin(\theta_1+\theta_2)}q_1\right)$$

$$\theta_1' = \begin{cases} \theta_1 & \lambda_1\cos\theta^* + \lambda_2\sin\theta^* \leq k\,(\lambda_1\cos\theta_1 + \lambda_2\sin\theta_1) \\ \theta & \lambda_1\cos\theta + \lambda_2\sin\theta > k\,(\lambda_1\cos\theta_1 + \lambda_2\sin\theta_1) \end{cases}$$

$$\theta_2' = \begin{cases} \theta_2 & \lambda_1\cos\theta + \lambda_2\sin\theta \leq k\,(\lambda_1\cos\theta_2 + \lambda_2\sin\theta_2) \\ \theta & \lambda_1\cos\theta + \lambda_2\sin\theta > k\,(\lambda_1\cos\theta_2 + \lambda_2\sin\theta_2) \end{cases}$$

[25] Other updating rules are conceivable. In a fast changing process it might be advisable for the competitors to require that a value overshoot for the optimal design seems to hold over a sequence of periods. One can also consider to tie the design changes to acquiring new equipment, and hence to the input of capital. Capital then charges a cost in the period of investment, and reinvestment is due when capital wears out. In such a model we cannot use the Cobb-Douglas function in (5.30). We rather need to choose some CES form the selection presented in Chap. 2.

[26] Note again that we speak interchangeably of quantity and price. This was discussed at length in connection with setting up Amoroso's formulas. When the commodities are nonhomogeneous then each competitor actually has a monopoly market. It is logical from Bertrand's argument to consider that the competitors set prices. However, in a monopoly, price and quantity are equivalent for optimization, and the convention from Amoroso is to use quantity. In this way we recognize standard economics textbook stuff. This, however, is far from mixing up the Cournot and Bertrand cases as we sometimes see. Competing commodities are either homogenous, or they are different. There is no other alternative, and it is the consumers who decide which is the case.

where

$$\lambda_1' = \frac{1}{2}\frac{1}{q_1 \cos\theta_1 + q_2 \cos\theta_2}$$

$$\lambda_2' = \frac{1}{2}\frac{1}{q_1 \sin\theta_1 + q_2 \sin\theta_2}$$

$$\theta' = \arctan\frac{\lambda_2}{\lambda_1}$$

References

Agliari A, Gardini L, Puu T (2005) Global bifurcations in duopoly when the Cournot point is destabilized through a subcritical Neimark bifurcation. Int Game Theory Rev 8:1–20

Bertrand J (1883) Théorie mathématique de la richesse sociale. Journal des Savants 48:499–508

Chamberlin EH (1933) The theory of monopolistic competition. A reorientation of the theory of value. Harvard University Press, Cambridge, MA

Cournot A (1838) Récherces sur les principes mathématiques de la théorie des richesses. Dunod, Paris

Edgeworth FY (1897) La teoria pura del monopolio. Giornale degli Economisti 15:13–31

Frisch R (1965) Theory of production. D. Reidel Publishing Company, Dordrecht, Holland (Translation from Norwegian of "Innledning til Produksjonsteorien" 1927, various editions.)

Hicks JR (1947) Value and capital. Oxford University Press, Oxford

Johansen L (1972) Production functions. North-Holland, Amsterdam

Lancaster KJ (1966) A new approach to consumer demand. J Polit Econ 74:132–157

Puu T (2017) A new approach to modeling Bertrand duopoly. Rev Behav Econ 4:51–67

Tramontana F, Gardini L, Puu T (2010) Global bifurcations in a piecewise smooth Cournot duopoly. Chaos Solitons Fractals 43:15–23

von Stackelberg H (1934) Marktform und Gleichgewicht. Julius Springer, Berlin

Chapter 6
Stackelberg

Abstract A Century after Cournot von Stackelberg proposed a modification of the model. He suggested that any duopolist may learn the competitor's Cournot reaction function, and so search for a better solution as a "leader". This, of course, works provided the other, the "follower", actually adheres to the proper reaction function. In this way two feasible leader/follower pairs can be formed. Both can also attempt leadership, in which case there would be economic warfare, from which presumably the firm with better financial backing might come out as winner. It was proved that Stackelberg equilibrium always is more profitable for the leader than Cournot equilibrium. Stackelberg never thought in dynamic terms—he could, for instance, have considered a variable supply policy over time to check if it might render a higher profit than a constant supply ∩ over time, but he did not. If we want to combine the static Stackelberg case with the dynamic Cournot process, things become really interesting. There may even seem to be a contradiction buried here. Stackelberg equilibrium is more profitable, no question about that, but, even the successful Stackelberg leader will ripe a higher temporary profit from jumping to Cournot action. This is so because the Cournot action was defined as the best move in any situation. However, this will not last for long—if the Cournot process is stable, then the orbit will approach Cournot equilibrium—and that is worse than Stackelberg leadership. So the stage is set for designing a rule for switching between Cournot and Stackelberg actions, which seems not to have been attempted before. The natural format is to let the competitors switch to Stackelberg leadership whenever the resulting profit is higher than, not the expected profit from Cournot action, because it never is, but, say, exceeds a fraction, say 50 or 75% of it. Incorporating such a rule of switching results in interesting scenarios, including both duopolists choosing Stackelberg leadership or both using Cournot best reply. The dynamics shows different periodic scenarios, multiplicity of attractors, and bifurcations. Some results are really unexpected, for instance that it may be more profitable than any other action to be a follower if the other becomes a leader, which, of course is not a choice under the follower's control.

T. Puu, *Disequilibrium Economics*, https://doi.org/10.1007/978-3-319-74415-5_6

6.1 Some History

6.1.1 Biographical

Baron Heinrich von Stackelberg, an offspring of the landowning German nobility in Estonia, then part of the Romanov Empire, was born in Moscow in 1905. He studied in Cologne and Berlin.

His doctoral theses dealt with the pure theory of production, and his "Habilitation" (complementary work to qualify as "Privatdozent", i.e. Associate Professor) dealt with duopoly. This is what he is remembered for.

He survived the War, but died during a business visit to Madrid in 1945, aged 40. He was among those German scientists who remained in the country and even joined the Nazis, though he is said to have left the party early when his relatives pointed out that it was not becoming for a nobleman to join such populace. It is tempting to compare him to location theorist August Lösch, born 1906, who also died in 1945. The latter even refused to give oath of fidelity to the Führer, compulsory for civil servants in Nazi Germany, and was therefore denied the permit to teach, the "Venia Legendi". Lösch lived as private scholar at the verge of starvation, and died from consequences of malnutrition.

6.1.2 Stackelberg Leadership

Stackelberg's idea was that the competitors can learn each other's Cournot reaction function and take account of it in designing their own reaction—becoming "leaders", thereby getting a higher equilibrium profit.

This is feasible, given the other actually adheres to the Cournot reaction, is a "follower" and does not attempt to become a leader as well. Stackelberg identified four cases—two different consistent leader/follower pairs. Further, both competitors attempting leadership, which is not consistent as then no competitor adheres to Cournot reaction under which the leadership action was calculated.[1] Further there is the case where both competitors act as followers, which is the Cournot case.

As Stackelberg's model subsumes Cournot's as one of four cases, it may be considered as a generalization. In another sense, however, Stackelberg is more restricted than Cournot: There is no idea of evolutionary dynamics. All is static equilibrium—true "game theory".[2]

[1] It has been argued that the realization that there can only be one leader in a stable system was the cause for Stackelberg becoming a "follower" of the "Führer", but to the present author this has the flavour of a too simplistic construct. We have no reason to believe that Stackelberg was that naïve—there were more substantial benefits from joining the political leadership.

[2] It is tempting to generalize the Stackelberg model from duopoly to oligopoly, for instance triopoly. This brings possibilities to define different levels of leadership, for instance being an intermediate

It is curious that Stackelberg never thought in dynamic terms: If one assumes that the follower really reacts mechanically to every move the leader makes, it is not farfetched to think that the leader might design a changing policy over time so as to drag the follower along and perhaps get an even higher average profit over time. Actually this seems to be possible in pure Stackelberg action, though the originator never thought of it.

We are not following that track; rather we will try to incorporate Stackelberg action in a dynamic Cournot model.[3] A pure dynamic Stackelberg model was discussed in Puu (2010). The combination model itself was first proposed in Puu (2009). There is a latent challenge in this: It has been proved that, considering equilibria alone, Stackelberg leadership always yields a higher profit for the leader than does a Cournot equilibrium. On the other hand, the best reply calculated under Cournot action, always returns the highest short run profit—even for the successful Stackelberg leader! This is because the Cournot action was defined and calculated this way.

So, one may imagine a Stackelberg leader who jumps out from leadership to get higher temporary profit. But, if the Cournot process is stable, then it starts converging and at some point very soon, it will be better to jump back to Stackelberg leadership. This sets the stage for a dynamic system.

6.1.3 The Stackelberg Model

To show his point Stackelberg assumes the somewhat curious (inverse) demand function[4]:

$$p = \exp\left(-\sqrt{q_1 + q_2}\right) \tag{6.1}$$

where p is price and q_1, q_2 denote the supplies of the competing duopolists.

The function corresponds to general intuition that demand decreases with increasing price, though it is ad hoc as it does not result from any known utility maximization. One may wonder about the square root as just $(q_1 + q_2)$ is simpler and renders the same qualitative features. The reason is obvious if one works through the case: Without the square root sign leadership action coincides with Cournot action, so he would not have been able to make his point! Further, the

leader, a follower to someone but leader to another. There have been published contributions in this direction, even by the present author. But it seems difficult to achieve anything really interesting. All tends to result in a sterile taxonomy of cases.

[3]In one of the stubs we propose a dynamic version set up as a dynamic programming problem. It is immediately apparent that the optimality conditions are not satisfied by the stationary conditions for traditional Stackelberg leadership. How much the stationary profits fall short of the dynamic remains a research agenda.

[4]Stackelberg also uses linear demand as an alternative, but linearity has its severe snags and limitations, so the exponential case is potentially more interesting.

variable production costs are assumed zero, so the competitors maximize just revenue,

$$R_1 = q_1 \exp\left(-\sqrt{q_1 + q_2}\right) \tag{6.2}$$

$$R_2 = q_2 \exp\left(-\sqrt{q_1 + q_2}\right) \tag{6.3}$$

respectively.

Without the zero cost assumption one could not solve for the explicit reaction functions.[5] In the model proposed below, we will rather take the isoelastic demand function, $p = (q_1 + q_2)^{-1}$ which, as we have seen, results from utility maximization, and yields explicit reaction functions even with nonzero costs.

6.1.4 Digression

It is, however, instructive to follow Stackelberg's original reasoning as it sheds light on our following discussion, and the reader is encouraged to verify the details.

Maximizing profits (=revenue) $R_1$the first competitor obtains the Cournot reaction function

$$q_1 = 2 + 2\sqrt{1 + q_2} \tag{6.4}$$

Due to symmetry, the second, maximizing R_2 likewise gets

$$q_2 = 2 + 2\sqrt{1 + q_1} \tag{6.5}$$

In the Cournot equilibrium point, intersection of the two reaction functions, the coordinates are readily obtained as $q_1 = q_2 = 8$. We can also calculate the profits in this point, equal for both, as $8\exp(-4) \approx 0.146$.

Now suppose the first competitor considers becoming a leader, taking account of the reaction function (6.5) $q_2 = 2 + 2\sqrt{1 + q_1}$.

The leader's profits, thus being,

$$R_1 = q_1 \exp\left(-\left(q_1 + 2 + 2\sqrt{1 + q_1}\right)\right) \tag{6.6}$$

render a maximum when the total (no longer partial) derivative is put equal to zero, which yields

$$q_1 = 2 + 2\sqrt{2} \approx 4.828$$

[5]The assumption of zero production costs is not an oddity only due to Stackelberg—it was used by many other good economists, such as Hotelling, Palander, Wald, and even Cournot himself. Occasionally the production of natural mineral water was proposed—as if the bottling were costless. Not quite convincing!

Using the proper Cournot reaction function, the second competitor (follower) then supplies

$$q_2 = 4 + 2\sqrt{2} \approx 6.828$$

The leader reduces supply to almost half of that for the Cournot case, the follower reduces as well, though not quite as much. As a consequence total market supply is reduced which drives the price up.

As we know the leader/follower coordinates, we can calculate the profits of the competitors as

$$R_1 = \left(2 + 2\sqrt{2}\right) \exp\left(-\sqrt{6 + 4\sqrt{2}}\right) \approx 0.159$$

and

$$R_2 = \left(4 + 2\sqrt{2}\right) \exp\left(-\sqrt{6 + 4\sqrt{2}}\right) \approx 0.225$$

respectively.

As proved in general, leadership always results in higher profits. But, the follower profits even more from the increased price, due to its larger supply.

This is no misprint! There is nothing to say that the follower must suffer a loss from leadership—sure someone has to lose from this reduction of competition, but it is the consumers who have to pay a higher price and suffer a reduction of consumer's surplus.

It could thus be even more profitable for the first competitor if the second attempted leadership. But, logically there is no way for a competitor to decide to be the (sole) follower.

Now comes a second surprise, which is crucial for the following modelling. Assume we start from a situation where the first competitor is the leader the second the follower. Then suppose the leader reverts to Cournot action—after all the reaction function is defined as the best reply in any situation—even for a successful duopoly leader.

Taking the follower's supply

$$q_2 = 4 + 2\sqrt{2} \approx 6.828$$

as given, the proper reaction function $q_1 = 2 + 2\sqrt{1 + q_2}$ (6.4) renders

$$q_1 = 2 + 2\sqrt{5 + 2\sqrt{2}} \approx 7.596$$

which is a substantial increase of supply, almost back to Cournot equilibrium. Substituting in the leader's profit expression $R_1 = q_1 \exp\left(-\sqrt{q_1 + q_2}\right)$ (6.6) (too long to be printed here) results in the approximate profit $R_1 \approx 0.170$, indeed higher than 0.159 in the leader position, and higher than the Cournot equilibrium profit 0.146.

But this only holds for the next move! The competitor will react to this, and if the process converges, the Cournot equilibrium will be the limit, which is definitely worse than Stackelberg leadership, and worse for the follower as well as we saw!

So, when does any of the competitors jump to Stackelberg action in a dynamic Cournot setting? Definitely not whenever the expected short run leadership profit is higher than what the Cournot reaction function prescribes, because it *never* is!

Nevertheless, a dynamic Cournot process with intermissions of leadership may result in higher average profits than either Cournot or Stackelberg equilibria.

The stage is set for dynamics, and the simplest clue seems to be to assume that reversion to leadership occurs whenever leadership profits exceeds some fraction (less than unity) of the expected Cournot reaction profit for the next move.

However, as mentioned, we will do this using a more convenient demand function. The present author does not think it is that fun with all Stackelberg's square roots.[6]

6.2 Unifying Cournot and Stackelberg

6.2.1 Agenda

We would like to formulate a single more general map in which both the Cournot and the Stackelberg equilibria can exist as fixed points. To this end we suggest a rule for flipping between Cournot and Stackelberg action on the basis of expected profits. The idea is to compare expected next period profits under Cournot action to profits resulting from Stackelberg leadership. But, as already mentioned, (expected) Cournot profits always exceed Stackelberg leadership profits, so we propose to locate the balance point between current Cournot and Stackelberg profits, by incorporating a special parameter.

By this, we can formulate a single model which contains both the Cournot and Stackelberg solutions as fixed points, even including the point where both competitors go on trying to be leaders.

[6]The reaction function of the competitor is not the only thing to learn. It may even seem easier to learn the outcome of the interactive dynamic, particularly if it is a low periodic orbit. This involves the inherent complication that learning periodicity and adapting is always bound to alter this very periodicity, to higher period and even more difficult to learn. This calls in doubt the economists' favourite idea of intertemporal equilibrium with perfect foresight of the future, now baptized with the absurd term "rational expectations". It seems that the fixed point is the only orbit about which one can have "rational expectation" and this brings us back to stationary situations and equilibrium theory. The problem is there with all the dynamic models, but we will only deal with it in the present chapter.

Having this agenda we choose a simple setup, used repeatedly by the present author: A smooth iso-elastic demand function combined with constant marginal costs for the competitors.

6.3 Model Setup

6.3.1 Cournot Action

Reaction Functions

Assume the inverse demand function

$$p = \frac{1}{q_1 + q_2} \tag{6.7}$$

where p denotes market price and q_1, q_2 denote the outputs of the duopolists. Given the competitors have constant marginal costs, denoted c_1, c_2 respectively, the profits are

$$V_1 = \frac{q_1}{q_1 + q_2} - c_1 q_1 \tag{6.8}$$

$$V_2 = \frac{q_2}{q_1 + q_2} - c_2 q_2 \tag{6.9}$$

Putting the derivatives to zero, $\partial V_1/\partial q_1 = 0$, $\partial V_2/\partial q_2 = 0$, and solving for q_1, q_2 one obtains the reaction functions

$$q_1' = \sqrt{\frac{q_2}{c_1}} - q_2 \tag{6.10}$$

$$q_2' = \sqrt{\frac{q_1}{c_2}} - q_1 \tag{6.11}$$

The dash, as usual, denotes the next iterate, i.e., the "best reply" of one competitor given the observed (=expected) supply of the other.

Constraints

Obviously, (6.10) returns a negative reply $q_1' < 0$ if $q_2 > 1/c_1$, and (6.11) a negative reply $q_2' < 0$ if $q_1 > 1/c_2$ To avoid this, we could put $q_1' = 0$ whenever $q_2 > 1/c_1$, and $q_2' = 0$ whenever $q_1 > 1/c_2$.

This means reformulating (6.10)–(6.11) as follows

$$q_1' = \begin{cases} \sqrt{\frac{q_2}{c_1}} - q_2 & q_2 \leq 1/c_1 \\ 0 & q_2 > 1/c_1 \end{cases} \tag{6.12}$$

$$q_2' = \begin{cases} \sqrt{\frac{q_1}{c2}} - q_1 & q_1 \leq 1/c_2 \\ 0 & q_1 > 1/c_2 \end{cases} \tag{6.13}$$

As we saw in Chap. 3, this map may involve the awkward case of orbits visiting the origin of phase space and sticking there, to the avoidance of which we introduced adaptation. However, we also saw that given the parameter ranges are restricted to $\frac{1}{2\sqrt{c_1 c_2}} - \frac{1}{4c_1} < \frac{1}{c_1}$ and $\frac{1}{2\sqrt{c_1 c_2}} - \frac{1}{4c_2} < \frac{1}{c_2}$, the orbits keep aloof of the zero branches, so, as we now want to focus other problems, we assume marginal costs to fulfil these constraints. The reader can easily make an adaptive version of the map, and study phenomena with a wider range of admissible parameter values.

However, as we learn from the numerical study below, jumping to Stackelberg leadership itself serves to stop the orbits from visiting negative branches of the reaction functions. In the sequel we see cases where the marginal cost ratio extends to $\frac{c_2}{c_1} \leq 16$ instead of being confined to the prescribed interval $\frac{c_2}{c_1} < 6.25$, yet without any problems with involvement of zero branches. However, as a mathematical study by Tramontana et al. (2010) demonstrates, the problems with negative (zero) branches still remain in the model version without adaptation.

Equilibrium

Putting $q_1' = q_1, q_2' = q_2$, one can as usual solve for the coordinates of the Cournot equilibrium point

$$q_1 = \frac{c_2}{(c_1 + c_2)^2} \tag{6.14}$$

$$q_2 = \frac{c_1}{(c_1 + c_2)^2} \tag{6.15}$$

Substituting back from (6.14)–(6.15) in (6.8)–(6.9), one gets the profits of the competitors in the Cournot equilibrium

$$V_1 = \frac{c_2^2}{(c_1 + c_2)^2} \tag{6.16}$$

$$V_2 = \frac{c_1^2}{(c_1 + c_2)^2} \tag{6.17}$$

It is obvious that the firm with lower unit costs obtains the higher profit.

6.3.2 Digression on Profits

Note that (6.16)–(6.17) are the profits in the Cournot *equilibrium* point. During the Cournot iteration process (6.10)–(6.11) profits can also be considerably higher than in equilibrium. As we will see below, the Stackelberg leadership profits are always higher than the Cournot equilibrium profits, but in the Cournot process temporary profits even exceed the leadership profits. To calculate temporary profits, just substitute from (6.10) into (6.8), and from (6.11) into (6.9), to obtain[7]

$$V_1 = \left(1 - \sqrt{c_1 q_2}\right)^2 \tag{6.18}$$

$$V_2 = \left(1 - \sqrt{c_2 q_1}\right)^2 \tag{6.19}$$

Stability

Of course, the Cournot equilibrium (6.14)–(6.15) can be stable or unstable. To check, calculate the derivatives of (6.10)–(6.11)

$$\frac{\partial q_1'}{\partial q_2} = \frac{1}{2}\frac{1}{\sqrt{c_1 q_2}} - 1 \tag{6.20}$$

$$\frac{\partial q_2'}{\partial q_1} = \frac{1}{2}\frac{1}{\sqrt{c_2 q_1}} - 1 \tag{6.21}$$

In the Cournot point, substitute from (6.14)–(6.15) to obtain

$$\frac{\partial q_1'}{\partial q_2} = \frac{1}{2}\left(\frac{c_2}{c_1} - 1\right) \tag{6.22}$$

$$\frac{\partial q_2'}{\partial q_1} = \frac{1}{2}\left(\frac{c_1}{c_2} - 1\right) \tag{6.23}$$

Note that both Cournot equilibrium derivatives have an infimum value $-\frac{1}{2}$. Hence neither can become smaller than -1, but either can become larger than 1, if $c_2 > 3c_1$ or $c_1 > 3c_2$. Also note that the derivatives always have opposite signs. To check stability, compose the Jacobian matrix

$$J = \begin{bmatrix} 0 & \frac{1}{2}\left(\frac{c_2}{c_1} - 1\right) \\ \frac{1}{2}\left(\frac{c_1}{c_2} - 1\right) & 0 \end{bmatrix} \tag{6.24}$$

[7]Note that these expressions may seem to be nonnegative. But if $q_2 > 1/c_1$ or $q_1 > 1/c_2$, i.e., the constraints for the first branches of (6.22)–(6.23) are violated, then this is due to the fact that negative costs dominate over negative revenues, which in terms of subject matter is nonsense. Anyhow, we already restricted the map (6.20)–(6.21) to (6.22)–(6.23), so we need not be further concerned.

If

$$|J| = \left| \frac{(c_1 - c_2)^2}{4c_1c_2} \right| < 1 \tag{6.25}$$

then the Cournot equilibrium is stable, otherwise not. The condition is equivalent to

$$\frac{c_2}{c_1} \in \left[3 - 2\sqrt{2}, 3 + 2\sqrt{2}\right] \tag{6.26}$$

6.3.3 Stackelberg Action

First Equilibrium

According to Stackelberg's idea, the competitor controlling q_1 can take the reaction function (6.11) of the other for given, and substitute it in its own proper profit function (6.8), to obtain

$$V_1 = \sqrt{c_1 q_1} - c_2 q_1 \tag{6.27}$$

Putting $dV_1/dq_1 = 0$, and solving, one gets

$$q_1 = \frac{c_2}{4c_1^2} \tag{6.28}$$

The corresponding value of q_2, provided that firm adheres to its Cournot reaction function, is obtained through substituting (6.28) in (6.11) and equals

$$q_2 = \frac{2c_1 - c_2}{4c_1^2} \tag{6.29}$$

Thus the followership output is positive only provided its marginal cost does not exceed twice that of the leader.

Using (6.28)–(6.29) in (6.9), the Stackelberg leadership profit can be easily calculated as

$$V_1 = \frac{c_2}{4c_1} \tag{6.30}$$

Second Equilibrium

Similarly, the second firm can try leadership, substituting (6.10) in (6.9), and maximizing

$$V_2 = \sqrt{c_2 q_2} - c_1 q_2 \tag{6.31}$$

to obtain

$$q_2 = \frac{c_1}{4c_2^2} \tag{6.32}$$

The corresponding Cournot response of the first firm would then be

$$q_1 = \frac{2c_2 - c_1}{4c_2^2} \tag{6.33}$$

Obviously both leaderships are possible as duopoly solutions (without the follower dropping out) provided $\frac{1}{2} < \frac{c_2}{c_1} < 2$ holds.

The Stackelberg leadership profit for the second firm is

$$V_2 = \frac{c_1}{4c_2} \tag{6.34}$$

It is easy to check, and has been proved many times, that Stackelberg leadership profits always exceed the respective Cournot *equilibrium* profits, i.e., that (6.30) is higher than (6.16), and (6.34) higher than (6.17).

As we saw with Stackelberg's own demand function, this does not prevent that the follower may profit even more.

Problems

These equilibrium solutions are consistent, provided one competitor in each case agrees to act as a Cournot follower. If not, there is a problem that was particularly focused by Stackelberg.

But there is more to it because Cournot's theory is dynamic, Stackelberg's static.

One may ask: Could the Stackelberg equilibria (6.28)–(6.29) and (6.32)–(6.33) be fitted into the Cournot dynamical system (6.10)–(6.11) as occasionally attracting fixed points of the map (6.12)–(6.13)?

The answer is no! If the first competitor chooses to try leadership according to (6.28), and the other indeed follows according to (6.29), then, in the next move, retaining leadership by the first firm is no longer optimal under the dynamic Cournot map.

The leader could in fact obtain a higher profit through returning to Cournot action; after all, the Cournot best reply was defined that way. The Stackelberg equilibria hence cannot be attracting fixed points of the usual dynamic Cournot map; the map has to be modified in order to accommodate these additional fixed points.

6.4 A Proposed Map

6.4.1 Profit Considerations

Above in (6.12)–(6.13), the constrained Cournot map was given. We now want to amend this map through considering possible jumping to Stackelberg leadership. It is close at hand then to compare current expected Cournot action profit, as given in (6.18)–(6.19), to Stackelberg leadership profit according to (6.30) and (6.34).

Assume that there is a certain proportionality constant $k < 1$ such that the firms keep to Cournot action as long as expected current profits are not less than Stackelberg leadership profits multiplied by this k, i.e. as long as,

$$\left(1 - \sqrt{c_1 q_2}\right)^2 \geq k \frac{c_2}{4c_1} \tag{6.35}$$

$$\left(1 - \sqrt{c_2 q_1}\right)^2 \geq k \frac{c_1}{4c_2} \tag{6.36}$$

These inequalities are more convenient to use in the maps if we solve for q_2, q_1:

$$q_2 \leq \left(1 - \sqrt{k \frac{c_2}{c_1}}\right)^2 \frac{1}{c_1} \tag{6.37}$$

$$q_1 \leq \left(1 - \sqrt{k \frac{c_1}{c_2}}\right)^2 \frac{1}{c_2} \tag{6.38}$$

The value of the parameter k indicates how willing the competitors are at attempting a jump to leadership action.

6.5 The Map

We can now specify the map resulting from these considerations with three branches for each actor.

$$q_1' = \begin{cases} \sqrt{\frac{q_2}{c_1}} - q_2 & q_2 \leq \left(1 - \sqrt{k \frac{c_2}{c_1}}\right)^2 \frac{1}{c_1} \\ \frac{c_2}{4c_1^2} & \left(1 - \sqrt{k \frac{c_2}{c_1}}\right)^2 \frac{1}{c_1} < q_2 \leq \frac{1}{c_1} \\ 0 & \frac{1}{c_1} < q_2 \end{cases} \tag{6.39}$$

$$q_2' = \begin{cases} \sqrt{\frac{q_1}{c_2}} - q_1 & q_1 \leq \left(1 - \sqrt{k \frac{c_1}{c_2}}\right)^2 \frac{1}{c_2} \\ \frac{c_2}{4c_1^2} & \left(1 - \sqrt{k \frac{c_1}{c_2}}\right)^2 \frac{1}{c_2} < q_1 \leq \frac{1}{c_2} \\ 0 & \frac{1}{c_2} < q_1 \end{cases} \tag{6.40}$$

This map can now accommodate the Cournot equilibrium as well as both Stackelberg equilibria.

Note that we only need the third branches in case $\left(1-\sqrt{k\frac{c_2}{c_1}}\right)^2 < 1$ or $\left(1-\sqrt{k\frac{c_1}{c_2}}\right)^2 < 1$ respectively. When any of these inequalities holds, then the intervals for the second branches in (6.39)–(6.40) do not exist as the lower bounds are higher than the upper bounds. Obviously this happens when $k > 4\frac{c_1}{c_2}$ or $k > 4\frac{c_2}{c_1}$.[8] In this chapter we will avoid these complications by constraining numerical phase plane experiment to moderate parameter values. In the displays of attractors and basins in Figs. 6.5, 6.6, 6.7, 6.8, 6.9, 6.10, and 6.11 we keep the cost ratio to $\frac{c_2}{c_1} = 3$, so the inequalities state $k > \frac{4}{3}$ and $k > 12$.

We look at the bifurcation series occurring when $0.60 \leq k \leq 0.75$ so the system is safely remote from this complication, though the issue remains an interesting research topic.

6.5.1 The Parameter Plane

Some issues in the parameter plane concerning existence and stability can be dealt with by analytical considerations, other must have recourse to computer aided numerical study. The results of both are displayed in Fig. 6.1.

Existence of Cournot Equilibrium

In order to confirm that the Cournot equilibrium point is indeed located in its proper region of application we have to check (6.37) and (6.38) for the Cournot point (6.14)–(6.15).

Substitutions result in just one single condition

$$k \leq \frac{4c_1c_2}{(c_1+c_2)^2} \tag{6.41}$$

which depends on k and the ratio $\frac{c_2}{c_1}$. The Cournot equilibrium (6.14)–(6.15) exists in its region of definition iff (6.41) holds. As the right hand side attains a unitary maximum value for $c_1 = c_2$, we conclude that for (6.41) to hold it is necessary that $k \leq 1$, which was our first intuitive assumption already when we formulated the map.

[8]Incidentally, there is no substantial reason to choose the same $k = k_1 = k_2$ for the two competitors. We just wanted to display bifurcations in the parameter plane in 2D, and, as c_1, c_2 only influence the system through their ratio, there remains just one dimension for k_1, k_2.

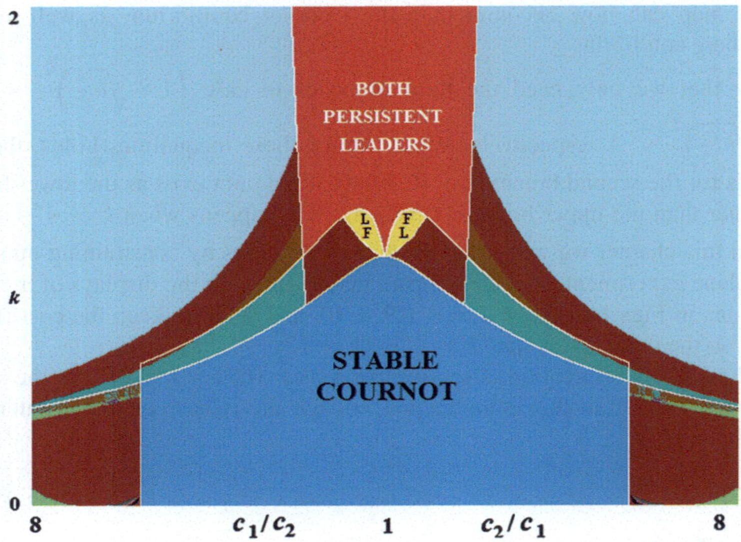

Fig. 6.1 Existence regions in parameter space for Cournot point (lower blue hump), the regular Stackelberg leader/follower pairs (small yellow lens shaped areas), and simultaneous persistence on Stackelberg leadership (upper red region with cuspoid bottomline)

Stability

The stability conditions for Cournot equilibrium were already stated above, i.e.,

$$\frac{c_2}{c_1} \in \left[3 - 2\sqrt{2}, 3 + 2\sqrt{2}\right] \tag{6.42}$$

As Cournot equilibria according to (6.41) can be defined for $\frac{c_2}{c_1} \in [0, \infty]$ the requirement on stability constrains the Cournot region.[9]

Stackelberg Equilibria

Existence

First Stackelberg Equilibrium In a similar way we can deal with the existence of the Stackelberg fixed points. Assume first that the first firm is the leader, the

[9]In terms of what we see in Fig. 6.1 the bounds are the vertical lines left and right in the picture. Note that when $\frac{c_2}{c_1} = 3 + 2\sqrt{2}$ then $\frac{c_1}{c_2} = 3 - 2\sqrt{2}$. This is so because $\left(3 - 2\sqrt{2}\right)\left(3 + 2\sqrt{2}\right) = 9 - 8 = 1$.

second the follower. Then the coordinates of the fixed point are (6.28)–(6.29), whereas the conditions for the branch definition are (6.35), sign reversed, and (6.36). Substituting from (6.28)–(6.29), we get two conditions,

$$k > \left(2 - \sqrt{2 - \frac{c_2}{c_1}}\right)^2 \frac{c_1}{c_2} \tag{6.43}$$

$$k \leq \left(2 - \frac{c_2}{c_1}\right)^2 \frac{c_2}{c_1} \tag{6.44}$$

These only depend on the parameter k and on the ratio $\frac{c_2}{c_1}$.

Second Stackelberg Equilibrium Reversing the roles of the firms, consider the fixed point (6.32)–(6.33), which must fulfil the branch conditions (6.35), and (6.36), sign reversed. Again, substitution gives

$$k \leq \left(2 - \frac{c_1}{c_2}\right)^2 \frac{c_1}{c_2} \tag{6.45}$$

$$k > \left(2 - \sqrt{2 - \frac{c_1}{c_2}}\right)^2 \frac{c_2}{c_1} \tag{6.46}$$

In parameter space $\frac{c_2}{c_1}, k$ (displayed in Fig. 6.1), the Stackelberg equilibria are located in their proper branch ranges in the two small yellow lens shaped areas.

Note that they are disjoint, and separated by a vertical line at $\frac{c_2}{c_1} = 1$. Hence, for each parameter combination, only one of the Stackelberg equilibria exists as a fixed point (the one where the firm with lower unit cost is the leader).

In the same picture we also display the lower region of existent Cournot fixed points. Note that it is disjoint from both Stackelberg equilibrium areas.

Persistence on Leadership by Both Competitors Finally, under the proposed map, there can also exist a fixed point where both firms keep to Stackelberg action, choosing (6.28) and (6.32) respectively. It is a fixed point of the map provided (6.35) and (6.36), signs reversed, hold, which upon substitutions from (6.28) and (6.32) become

$$k > \left(2 - \frac{c_1}{c_2}\right)^2 \frac{c_1}{c_2} \tag{6.47}$$

$$k > \left(2 - \frac{c_2}{c_1}\right)^2 \frac{c_2}{c_1} \tag{6.48}$$

The region in Fig. 6.1 of the parameter plane with the saw-toothed lower boundary represents the cases where both firms keep to Stackelberg leadership, though it is not any normal Stackelberg equilibrium. (In addition we have two more fixed points, where any one firm persists at its Stackelberg leadership, whereas the other firm drops out. Their regions of definition are not represented in the picture.)

As for the region outside the depicted existence regions for the fixed points, it is not possible to say from the preceding simple analysis which kinds of attractors may emerge there. Numerical experiment indicates periodic orbits.

These fixed point regions were calculated using analytic boundary curves from the five inequalities discussed in this section, (6.41) and (6.43)–(6.46). Then the areas were filled with appropriate colour for the attracting sets. The construction curves to which the numerical results fitted like hand in glove where then erased.

Note further the way Fig. 6.1 is constructed. The system is completely symmetric with respect to $\frac{c_2}{c_1}$ and $\frac{c_1}{c_2}$. Therefore, actually two diagrams, in $\frac{c_2}{c_1}$, k-space and in $\frac{c_1}{c_2}$, k-space respectively were put back to back, with $\frac{c_2}{c_1}$ increasing to the right, $\frac{c_1}{c_2}$ increasing to the left and $\frac{c_2}{c_1} = \frac{c_1}{c_2} = 1$ as the common dividing line. Had we chosen just one of the ratios as the horizontal coordinate, then one half of the diagram would have been squeezed in the interval $(0, 1]$ whereas the other would extend over $[1, \infty)$. This distortion would make it impossible to see the symmetry.

Stability

We already concluded that the regular Stackelberg equilibria are stable if

$$\frac{1}{2} < \frac{c_2}{c_1} < 2 \tag{6.49}$$

Accordingly, the Stackelberg equilibria, unlike the Cournot equilibria, are always stable when they exist.

6.6 Numerics

6.6.1 Parameter Space: Attractors and Bifurcations

Fixed Points

In Fig. 6.2, the bifurcation diagram, resulting from numerical experiment performed to detect periodic orbits of the system (6.39)–(6.40) when starting from a point close to the Cournot equilibrium point, is shown. To improve resolution, only the right half of the diagram as compared to Fig. 6.1 is displayed.[10] Obviously most of the plane contains areas representing period 1 orbits, i.e., attracting fixed points. These are of four different types (six in the format of Fig. 6.1), most of which can also be seen in Fig. 6.1 that represented existence regions for fixed points.

[10]We also kept the five construction curves for the fixed point regions this time.

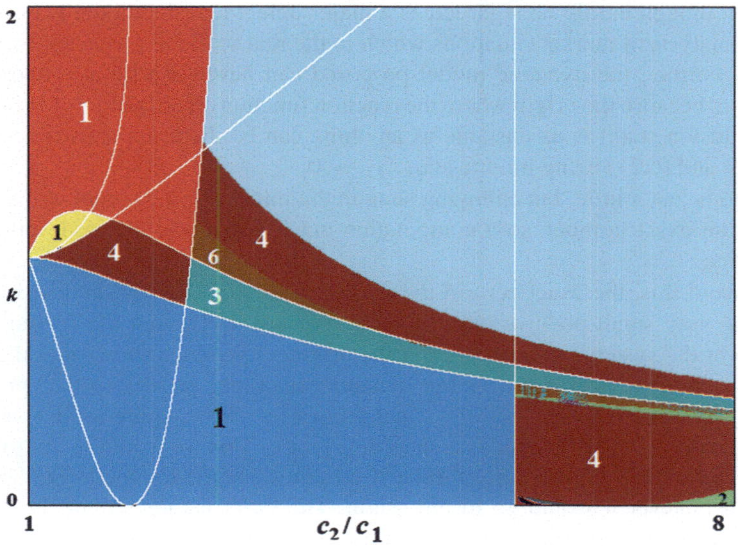

Fig. 6.2 Detailed picture of the right half of Fig. 6.1. In addition to the various fixed points shown in Fig. 6.1, we see periodic orbits of periods 2, 3, 4, and 6. In the right half of the picture we can see some smeared out colour, indicating coexistence of periods 2, 3, and 4

1. At the bottom of the picture the Cournot attracting equilibrium area is shown. It looks quite as in Fig. 6.1, though it yields to periodic orbits where the Cournot equilibrium turns unstable, i.e., at $\frac{c_2}{c_1} = 3 + 2\sqrt{2} \approx 5.83$, as we saw.
2. There is also the tiny yellow lemon shaped attracting Stackelberg area for the leader/follower pair, also quite as in Fig. 6.1.
3. Likewise there is the red region with the cuspoid bottom line, representing the non-standard Stackelberg case where both competitors insist on sticking to Stackelberg leadership (which Stackelberg in the equilibrium format considered unresolvable). True, both competitors get disappointed, as the assumption that the other acts as a follower does not hold, but under the defined map, neither could find expected current profits from Cournot action more profitable. Also this region has a corresponding one in Fig. 6.1.
4. There is further the light blue region in the upper right part of Fig. 6.2 which represents the case where the first competitor sticks to Stackelberg leadership, whereas the second finds no better action than to suppress production altogether. It might seem to be a kind of monopoly. However, as mentioned in the introduction, the model, in particular the iso-elastic demand function, is not suitable to treat either monopoly or collusion.[11]

[11]The reason is that under iso-elastic demand market revenue is a constant. Hence a monopolist (or a pair of collusive duopolists) could retrieve the whole revenue as profit without incurring any production costs if they produce nothing and sell this nothing at an infinite price. Such a solution is purely mathematical and has no meaning in terms of economics; it is just a shortcoming of one assumption.

It can though hardly be regarded as a major defect that the model is unsuitable for the analysis of market situations which in the real world as a rule are forbidden by law. Further, the dynamic model proposed can never end up at monopoly or collusion, because the origin where the reaction functions (6.10) and (6.11) intersect (the collusion state) is as unstable as anything can be, both derivatives according to (6.20) and (6.21) being infinite at $q_1, q_2 \rightarrow 0$.

There is just a little, but annoying snag in the numerics; the computer interprets zero as an exact number, so a computation may stick to the origin despite of its instability.

To avoid this, the exact zero in the definition of (6.39)–(6.40) can be replaced by some very small positive number ϵ in experiment. Though the reason is just to prevent the system from sticking to the origin, experience shows that the exact size of ϵ becomes essential for the dynamic outcome. An $\epsilon = 10^{-6}$ produces orbits different from an $\epsilon = 10^{-5}$ and from $\epsilon = 10^{-7}$, quite as if ϵ were an essential parameter of the model such as c_1, c_2, or k. This we would prefer to avoid. Unfortunately, the issue is quite similar if we add some random perturbation to the system whenever it would go to the origin. The exact parameters of the random number generator again end up as essential elements of the dynamic system. The reader might check this out, for the present model, or for the original Cournot model.

Therefore, the present author has come to the conviction that it is better to put these models, and all similar in an adaptive format, i.e., assume that the reactions do not immediately jump to the best reply, but only go some fraction >0 and <1 of the way from the previous move to next calculated best reply. In this way the system never goes to the origin—if the calculated best reply is negative, and one replaces it by zero, then the reaction only goes a fraction of the way towards the origin. In this way the system never goes to the problematic origin in finite time.

The adaptive format also adds interesting new features to the model, converting it from a diagonal format $q_1' = \phi_1(q_2)$, $q_2' = \phi_2(q_1)$ to the more general $q_1' = \phi_1(q_1, q_2)$, $q_2' = \phi_2(q_1, q_2)$. We do not pursue this issue here, but leave it as a project, outlined at the end of this chapter.

This laudatio for adaptation may seem annoying as it is repeated in other chapters, though we chose to include it for readers who are only interested in this particular model, the readers of the entire book can skip these paragraphs.

Periodic Orbits

As so much of the area in Fig. 6.2 is taken up by fixed point regions, little is left for other attracting orbits.

To increase resolution we blew up the interesting region in Fig. 6.3, in which we can see periodic orbits, labelled 3, 4, 8, 10, 12, and 14.

The remaining black area may contain higher periodicities, quasiperiodicity, or even chaos.

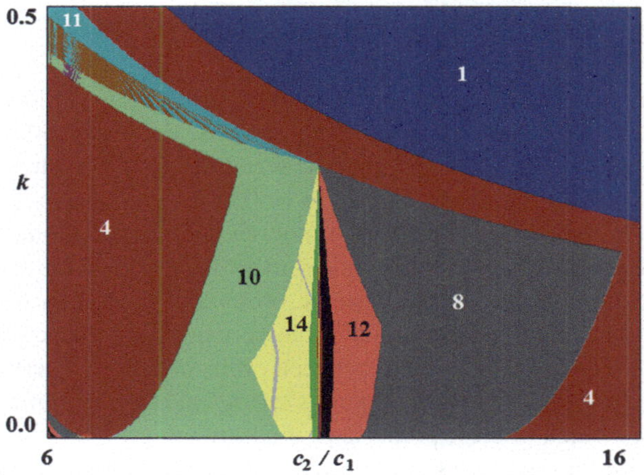

Fig. 6.3 Closer view of an interesting region in parameter space. Except the fixed point region, we find orbits of periods 4, 8, 10, 11, 12, and 14. In the lower left corner there are regions too small to be clearly seen which call for further enlargement. Further the black area indicates higher periodicity, quasiperiodicity, or chaos. Also note the smeared out colours on the upper left indicating coexistence of periods 6, 10, and 11

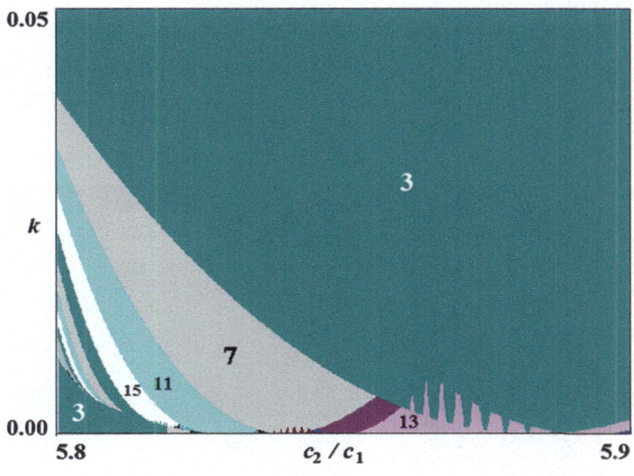

Fig. 6.4 Further close up picture of lower left corner of Fig. 6.3 to show more of periodic orbits and overlaps. The black regions again indicate higher periodicities or more complex dynamics

Figure 6.4 is a further blow up in the interval $\frac{c_2}{c_1} \in (6.8, 6.9)$, and $k < 0.05$. Notable are the two humps of a spiky appearance. They indicate coexistence of attractors; for some parameter combinations the initially chosen start condition sometimes leads to one periodicity, sometimes to another. As we will see below, the model is proficient in such coexisting attractors.

6.6.2 *The Phase Plane: Attractors and Basins*

There are many potentially interesting bifurcation scenarios to explore. We will just pick one simple series, with $c_1 = 0.5$, $c_2 = 1.5$ fixed, and $k \in [0.60, 0.75]$.

In Figs. 6.5, 6.6, 6.7, 6.8, and 6.9, the first series of phase diagrams are displayed for $k = 0.650$, $k = 0.700$, $k = 0, 745$, and $k = 0.750$.

The phase diagrams show the two Cournot reaction curves, along with the lines representing Stackelberg leadership action, further the different attractors, and their basins of attraction. Note that the periodic orbits are located on the intersection points of a grid of three horizontal and three vertical lines, hence producing nine attractor points. These are distributed between orbits of different periodicities, such as 1+ 4+ 4 =9. Further, *note that these points are independent of parameter k*, which only determines which of these points appear as attracting.

Figure 6.5 displays the case $k = 0.65$. The higher Stackelberg line is not visited, otherwise there are already the two coexistent 4-cycles, one SIM and one SEQ. Further, the Cournot point is stable too, within its now rather vast basin. In all, there are $1 + 4 + 4 = 9$ periodic points on a grid of three by three lines.

Following the Orbits

First, take a look at Fig. 6.6, where $k = 0.7$. As the fixed Cournot point according to (6.41) only ceases to exist at $k = 0.75$, it remains stable throughout the interval

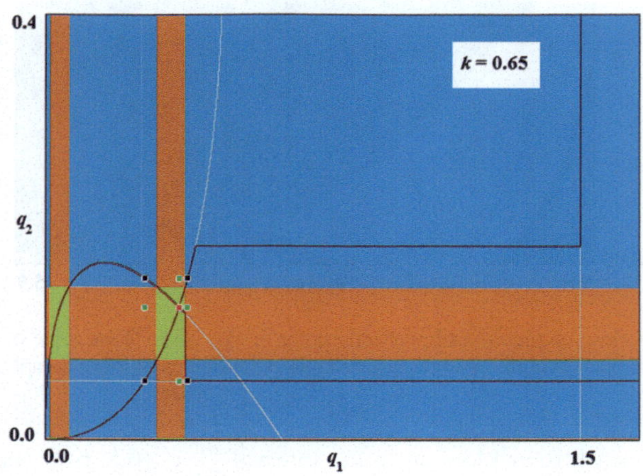

Fig. 6.5 Coexistent Cournot equilibrium point and two 4-period (SEQ and SIM) cycles when $c_1/c_2 = 3$, and $k = 0.65$. The SEQ cycle has the dominant blue basin of attraction, whereas the SIM cycle attracts from the orange "double cross" in this flag like picture. The Cournot equilibrium has the smallest basin, only the green rectangle and its preimage

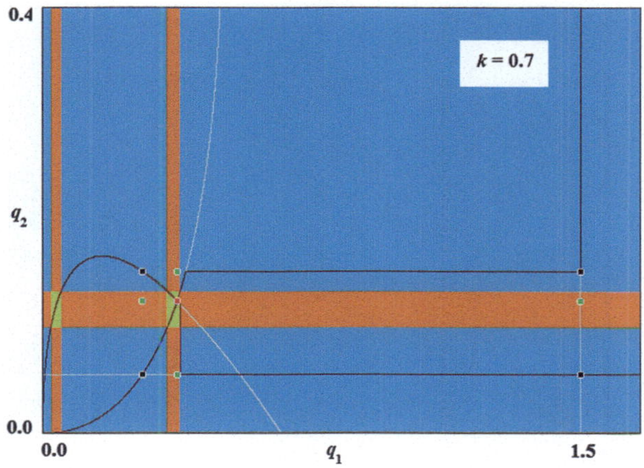

Fig. 6.6 Again coexistence of Cournot equilibrium with two 4-period SEQ and SIM cycles. Compared to Fig. 6.5 the case still represents $c_2/c_1 = 3$, but $k = 0.7$. has increased. There are two consequences of this. First, the orange cross and its green intersections have diminished in size, so the SEQ cycle is even more dominant than before. If this process of diminuation continues we expect an approaching bifurcation, which we indeed will see in the sequel. Second, the attractor points on the right now involve the Stackelberg line to the right, instead of the Cournot curves, the reason being that the transition point from Cournot to Stackelberg in the reaction functions has changed

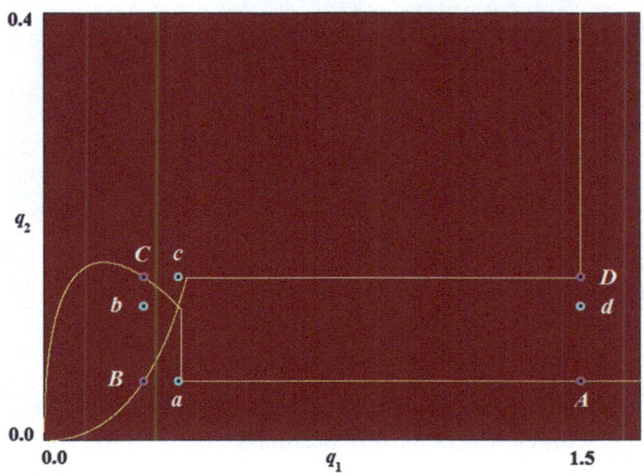

Fig. 6.7 Picture of the effective reaction functions (combinations of Cournot and Stackelberg) of the case shown in Fig. 6.6 when $k = 0.7$, for which we are going to calculate the coordinates of the attractor points. **A, B, C,** and **D** belong to the SEQ-cycle, **a, b, c,** and **d** to the SIM cycle

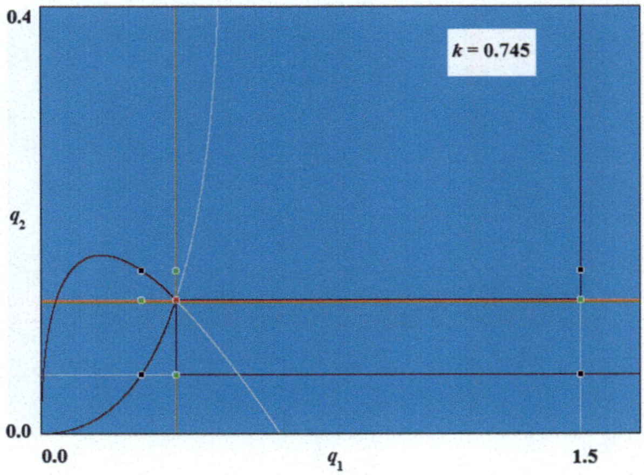

Fig. 6.8 Picture at $c_2/c_1 = 3$, with $k = 0.745$. As we see two basins are about to disappear

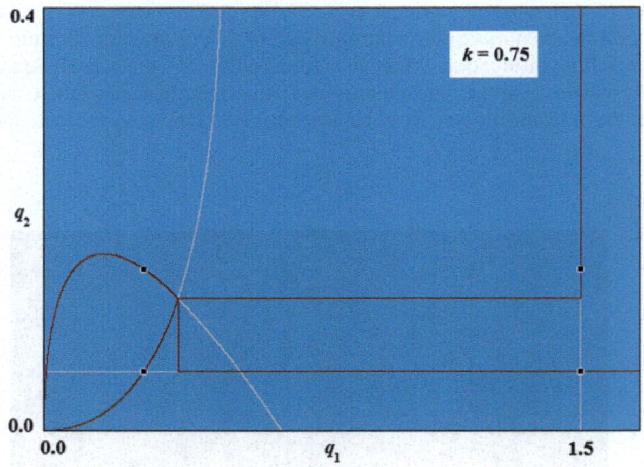

Fig. 6.9 The subcritical bifurcation, only the four-period SEQ cycle remains when $c_2/c_1 = 3$, and $k = 0.75$

studied. Its immediate basin is the tiny greenish rectangle around the intersection of the reaction curves, and a few other tiny preimages in the same shade.

In addition there is a 4-period orbit whose points lie on the Cournot reaction function (including the Stackelberg lines). Some researchers call such orbits "Markov perfect equilibria" (MPE). In the present author's opinion this is misleading, as it is a periodic orbit and no equilibrium point.

Earlier literature instead distinguished between "sequential adjustment" (SEQ), where the competitors take turns in adjustment every second period, and "simultaneous adjustment" (SIM), where both adjust at the same time every move. It was believed that the adjustment type must be established beforehand.

The points of a SEQ orbit necessarily lie on the reaction functions. Later it became obvious that the SIM system, which was more general, could itself also settle on a SEQ orbit.

Despite this mistake the present author finds it better to refer to the first 4-orbit as SIM, and to the second 4-orbit, with its points in the brownish cross, as SEQ.

Some of the 4-cycles are easy to follow analytically, for instance the case illustrated in Fig. 6.6, which flips between Cournot and Stackelberg action for both competitors. In the whole series of attractor/basin illustrations we keep to $c_1 = 0.5$ and $c_2 = 1.5$.[12] In Fig. 6.5 ($k = 0.70$), and unlike Fig. 6.5 ($k = 0.65$), the SEQ cycle also involves the Stackelberg lines.

Recall that it is only the costs c_1, c_2 that matter for the Cournot reaction curves and for the Stackelberg lines; therefore *the cycle points, and hence also the cycle profits, are completely independent of the value of* k. This parameter only has influence on which cycles exist as attractors in the dynamic process.

The SEQ Cycle

Let us so first calculate the points of the SEQ 4-cycle shown in Fig. 6.7. Start with B where firm 2 is a leader and firm 1 a follower. We know the coordinates from our general formulas,

$$q_1^B = \frac{2c_2 - c_1}{4c_2^2} = \frac{5}{18} \tag{6.50}$$

$$q_2^B = \frac{c_1}{4c_2^2} = \frac{1}{18}$$

In the next move (point C) firm 2 reverts from leadership to Cournot action. As it is a SEQ cycle we follow, firm 1 retains the previous move $q_1^C = q_1^B$, whereas $q_2^C = \sqrt{\frac{q_1^B}{c_2}} - q_1^B$.

Substituting for q_1^B we get

$$q_1^C = \frac{2c_2 - c_1}{4c_2^2} = \frac{5}{18} \tag{6.51}$$

$$q_2^C = \frac{1}{2c_2}\left(\sqrt{\frac{(2c_2 - c_1)}{c_2}} - \frac{1}{2}\frac{(2c_2 - c_1)}{c_2}\right) = \frac{2\sqrt{15} - 5}{18}$$

[12] As we know, it is only the ratio $\frac{c_2}{c_1} = 3$ that counts for the dynamics. The absolute values of unit production costs only have importance for the level of profits.

At next stage (point D), firm 1 chooses leadership, so the new coordinates become

$$q_1^D = \frac{c_2}{4c_1^2} = \frac{3}{2} \tag{6.52}$$

$$q_2^D = \frac{1}{2c_2}\left(\sqrt{\frac{(2c_2 - c_1)}{c_2}} - \frac{1}{2}\frac{(2c_2 - c_1)}{c_2}\right) = \frac{2\sqrt{15} - 5}{18}$$

Then, firm 2, next to respond, cannot choose Cournot follower action as the reaction function returns a negative coordinate $\frac{2c_1 - c_2}{4c_1^2} = -0.5$. However, it too can choose leadership.

So in the final move (point A), back to start, both become leaders:

$$q_1^A = \frac{c_2}{4c_1^2} = \frac{3}{2} \tag{6.53}$$

$$q_2^A = \frac{c_1}{4c_2^2} = \frac{1}{18}$$

We thus have the entire sequence of periodic points $A = \left(\frac{3}{2}, \frac{1}{18}\right)$, $B = \left(\frac{5}{18}, \frac{1}{18}\right)$, $C = \left(\frac{5}{18}, \frac{2\sqrt{15}-5}{18}\right)$, and $D = \left(\frac{3}{2}, \frac{2\sqrt{15}-5}{18}\right)$. The cycle starts out with both competitors being leaders. Next, firm 2 reverts to followership. Then firm 1 as well reverts to Cournot action. In the penultimate move firm 2 chooses leadership. And, finally, firm 1 as well chooses leadership. And so it goes for ever. In Fig. 6.7 we see how the points fit to the Cournot curves and the Stackelberg lines.

As we know all the coordinates, we could calculate the profits over a cycle, which is instructive to do. Much research remains in connection with investigating the average profits over cycles and remodelling with different learning assumptions in case the results are disappointing.

The SIM Cycle

We know from Fig. 6.6 that the SEQ cycle coexists with a SIM cycle, also of period 4, so we can also calculate its coordinates. In fact the latter has a much larger basin of attraction, as we see. A glance at Fig. 6.6 shows that all the attractor points locate on horizontal and vertical lines, which is rule in diagonal systems with $q_1' = \phi_1(q_2)$, $q_2' = \phi_2(q_1)$, where q_1 is missing as argument in ϕ_1, and q_2 in ϕ_2.

Thus, for each of the SIM points a, b, c, and d in Fig. 6.7 one coordinate coincides with one of those we already calculated for the SEQ cycle, whereas the other is one of the coordinates of the Cournot point. It is therefore useful to know the latter for our numerical case:

$$q_1^{\text{Cournot}} = \frac{c_2}{(c_1 + c_2)^2} = \frac{3}{8} \tag{6.54}$$
$$q_2^{\text{Cournot}} = \frac{c_1}{(c_1 + c_2)^2} = \frac{1}{8}$$

Now we are able to piece together the points of the SIM cycle, $a = \left(\frac{3}{8}, \frac{1}{18}\right)$, $b = \left(\frac{5}{18}, \frac{1}{8}\right)$, $c = \left(\frac{3}{8}, \frac{2\sqrt{15}-5}{18}\right)$, and $d = \left(\frac{3}{2}, \frac{1}{8}\right)$. Again the reader is advised to check the points using the iterative map (Cournot reaction, including the Stackelberg relapses when $k = 0.70$), to see how next point follows the previous, and also to calculate the average profit, to compare with Cournot and Stackelberg equilibrium profits.

The Cournot equilibrium also remains as a fixed point, not marked in Fig. 6.7, but as we see from Fig. 6.6, its basin only consists of a tiny rectangle and a few preimages. The two 4-period cycles and the Cournot fixed point produce 9 attractor points on a 3 by 3 lattice.

Profits

As we know the exact coordinates of the cycles and fixed points, we can easily calculate the profits using formulas (6.8)–(6.9). Substituting the coordinates of the SEQ cycle $ABCD$ in (6.8)–(6.9) and averaging over four periods we get

$$V_1 = 0.3932$$
$$V_2 = 0.0046$$

Compared to the Cournot equilibrium profits (6.54)

$$V_1 = 0.5625$$
$$V_2 = 0.0625$$

this is rather disappointing—for both competitors.

The SIM cycle yields a slightly better average, especially for the second competitor

$$V_1 = 0.4826$$
$$V_2 = 0.0296$$

though still not as good as the Cournot case.

Finally consider firm 2 as leader, firm 1 as follower in a Stackelberg point. Then

$$V_1 = 0.6944$$

$$V_2 = 0.0833$$

No doubt, this is the best solution for both competitors. Even the first firm, the follower, gains from the leadership of the other firm, quite as we saw using the original Stackelberg demand function.

For cycles of higher periods, we cannot calculate the orbit, but the computer can both trace the orbits and calculate average profits. This remains interesting research agenda.

Upper Bifurcation

Next, consider Fig. 6.8, where $k = 0.745$, and the brown cross with its green intersection has almost disappeared. The basin of the Cournot point has shrunk to almost nothing, as has the cross itself, which provides the basin for the 4-period SIM cycle. The same orbits remain as in the previous picture, though two of them (Cournot equilibrium and the SIM cycle) are almost about to disappear as attractors. Then the process goes to the SEQ cycle from almost all initial points in the blue shade basin.

This indeed happens at $k = 0.75$, the subcritical bifurcation shown in Fig. 6.9, where both the Cournot equilibrium, and the 4-period SIM cycle have actually disappeared.

What remains is one attractor, the 4-period SEQ cycle. In all, there now remain four grid points in the intersections of two (horizontal and vertical) lines. For further increasing k, only this cycle remains, until it is replaced by just one of its grid points, the intersection of the Stackelberg lines.

Subcriticality

The fact that Cournot point, contained in a small basin, remains along with other attractors indicates that the bifurcation from Cournot point to periodic orbits is subcritical, i.e., the fixed point is not just destabilized and replaced by another attractor as in the supercritical bifurcation. It rather disappears through its basin contracting and eventually vanishing, so that other attractors that already coexisted with it remain the only ones. The scenarios connected with such subcritical bifurcations often display quite complicated bifurcation structures in the boundary regions as shown in Agliari et al. (2005). Subcritical bifurcations are "hard" compared to the "soft" produced by supercritical bifurcations, and lead to dramatic changes in the system when they occur.

Lower Bifurcation

For k increasing from below, this case emerges at about $k \in [0.61, 0.62]$. For lower k, there is just the Cournot equilibrium point and one single basin as shown in Fig. 6.10.

In Fig. 6.11 we display the situation just after this lower bifurcation at about $k = 0.611$.

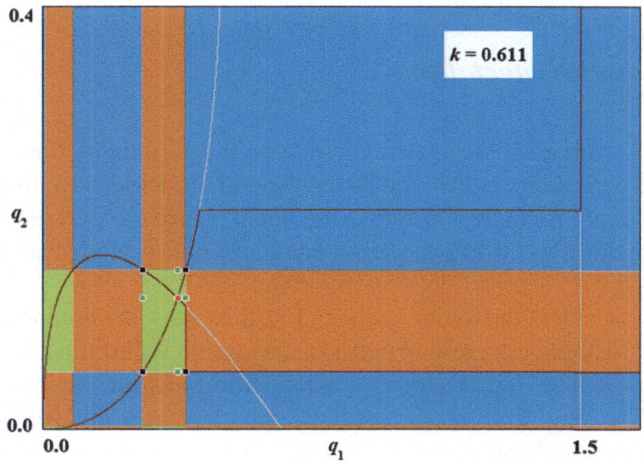

Fig. 6.10 Prior to the bifurcation where two new 4-period attractors are born, the system only has the Cournot point as attractor, and the whole phase space is covered by its greenish basin. The case reminds of Anna Agliari's studies (in which the present author had a little share) of the much more intricate bifurcations prior to the subcritical. If we put the system in an adaptive format, it is possible that we get the entire Agliari bifurcation series back

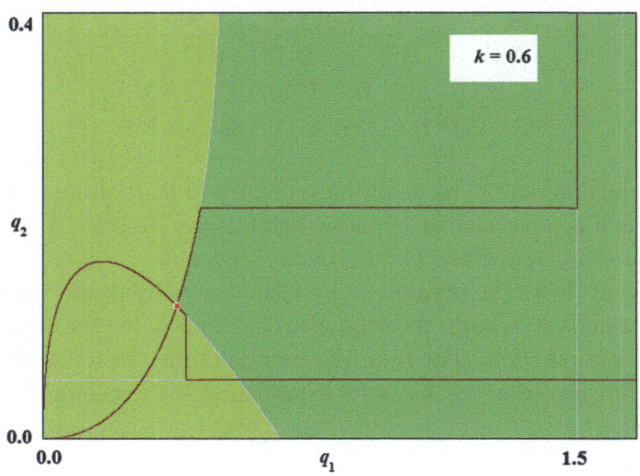

Fig. 6.11 The case just after the bifurcation in which the Cournot equilibrium is augmented by two 4-period cycles, one SEQ and one SIM. In this case the cross and its intersection are as large as they ever get in the picture series studied

There remains one bifurcation after the case of Fig. 6.8 where we had $k = 0.75$. In the picture we see that just one 4-cycle remains with two of its points on the Stackelberg lines. A further increase of k causes the transit point to Stackelberg action for competitor 2 to be lowered. Eventually it comes down to the Stackelberg line for firm 1. This happens when $k = \left(2 - \frac{c_1}{c_2}\right)^2 \frac{c_1}{c_2}$, or, with $c_1 = 0.5, c_2 = 1.5$, when $k = \frac{25}{27}$. Check this!

6.6.3 *Rational Expectations*

The question now arises if the agents could learn the periodicity they produce. This is easiest if the periodicity is low and the coexistent attractors have the same period.

Given the simple regularity, suppose the agents learn the periodicity, and react, not to the competitor's action one period back, but four periods back. What would be the outcome?

If one only considers the dynamic every fourth period, the outcome is exactly as before. But, in the three intervening periods, the system will settle to an independent cycle of the same type. The result is a composition of four 4-cycles successively displaced in time, and, even if we just have one original cycle, it is impossible to make the composition break down to a 4-cycle. The outcome is a 16-period cycle.

To see the point, consider a single four-cycle, whose points are denoted ABCD.

Try to arrange a four cycle. In the first line of the table there are three blanks between the entries. To produce ABCD again, choose BCDA for the first blanks, CDAB for the second, and DABC for the third, thus completing the sequence. (Each new choice of entries is indicated in bold face.)

A - - - **B** - - - **C** - - - **D** - - -
A **B** - - B **C** - - C **D** - - D **A** - -
A B **C** - B C **D** - C D **A** - D A **B** -
A B C **D** B C D **A** C D A **B** D A B **C**

So what is the outcome? The resulting sequence in the bottom row indeed starts with ABCD, but it is not repeated, because then follows BCDA, CDAB, and DABC. Only the full 16-sequence of the bottom line is repeated.

So, if the agents learn the periodicity of 4, they actually produce a periodicity of 16. The reader can try to arrange the sequences differently, but the outcome is always the same; a 16-period cycle. The first four entries in the sequence of 16 determine all the following, according to the order established in the original four-cycle, and as we have four choices for each entry, there would now seem to exist $4^4 = 256$ different cycles.

However, only 16 are different. The reason is as follows. Each recurrent sequence of 16 entries contains 16 different 4-sequences of entries, as we can start with any one of the entries. (If we start with one of the three last entries, we just have to add one, two, or three of the entries in the beginning of the full sequence.) Any four subsequent entries in the sequence determine the whole 16-period cycle, any of the 256 sequences belongs to a group of 16 identical orbits (only chosen with different starting points). Hence, the total number of distinct orbits is 16.

Note that average profits in the 16-period cycle remain the same as in the 4-period cycle, because the same points ABCD are visited with the same relative frequency as before though under different permutations.

To sum up; starting with one 4-period cycle and one basin, learning produced 16 different 16-period cycles, and, of course, 16 different basins, though. we cannot produce any picture of the basins in the phase diagram as it is 8-dimensional.

And so it goes on, learning and adjusting to the 16 periodicity produces a host of different 256-period cycles. The conclusion is that the idea of "rational expectations" is untenable even in the simplest case of only one original orbit of low order. The only periodicity that is possible to learn is a fixed point. For a general argument, see Puu (2006).

References

Puu T (2006) Rational expectations and the Cournot-Theocharis problem. Discret Dyn Nat Soc 32103:1–11

Puu T (2009) Unifying Cournot and Stackelberg duopoly in one dynamical system. In: Chiarella C, Bischi GI, Gardini L (eds) Nonlinear dynamics in economics and finance. Springer, Berlin, pp 117–137. ISBN 978-3-642-04022-1

Puu T (2010) Dynamics of Stackelberg duopoly. In: Puu T, Panchuk A (eds) Nonlinear economic dynamics. Nova Science Publishers, New York, pp 121–134. ISBN 978-1-61668-788-5

Tramontana F, Gardini L, Puu T (2010) Mathematical properties of a combined Cournot-Stackelberg model. Chaos Solitons Fractals 44:58–70

von Stackelberg H (1934) Marktform und Gleichgewicht. Springer, Berlin

von Stackelberg H (1938) Probleme der unvollkommenen Konkurrenz. Weltwirtschaftliches Arch 48:95

Chapter 7
Hotelling Duopoly

Abstract The Hotelling model was a very clever way to deal with Bertrand oligopoly. The commodities were considered as perfectly homogenous, but the suppliers were separated in geographical space and thus had local monopoly areas. Local price, mill price accrued with transportation costs was what mattered for the consumers and this made it possible to charge different mill prices for the same commodity. If the transportation cost was very high then the monopoly areas could be disjoint, and there would be no competition at all, otherwise they shared a market boundary which would provide for competition. A lowered mill price would decrease the proceeds per unit of product, but it would also increase the market area through pushing away the boundary from the mill's location. We are intentionally vague on space and boundary. One would naturally want the issue set in true 2-dimensional geographical space, though economists almost never managed. So, Hotelling's setting is a line interval in which the competitors locate—he refers to an intercontinental railway with stops at the locations, later vulgarized to ice-cream vendors on a shore. Well 2D is difficult, so we should not blame him too much. Worth blame is, however, that he assumed demand to be inelastic. Consumers would choose the cheapest supplier—mill price plus accrued transportation charge—but once this was done they would buy just one unit no matter what the price was. This seems contradictory. As location choice is free, Hotelling concludes that the competitors both crowd in the centre of the interval, and then we are back at the mess of Bertrand's original case. The cleverly introduced space goes out again. In a short passage Hotelling notes that if demand were elastic then this extreme crowding would not occur, the competitors would locate apart, not at the quartiles as would be socially optimal, but somewhat closer due to the contested middle area. It is a mystery that Hotelling did not make this his main case. Everybody was thrilled at the paradox of crowding in the middle, and the extremely popular model was interpreted in different ways by analogy—similarity of competing goods, convergence to the centre of political ideologies, and whatever in terms of "variables" not defined and not even measurable. Only a few years after Hotelling's article Lerner and Singer in an ingenious graphical analysis showed that the crowding did not occur, even when the consumers just had a reservation price. This means they would buy one unit if local price did not exceed the reservation price, otherwise they would buy nothing.

T. Puu, *Disequilibrium Economics*, https://doi.org/10.1007/978-3-319-74415-5_7

Around 1940 Smithies set up the entire model with linear demand functions, but he claimed that the revenue integrals were too difficult to evaluate. They were not, rather quite simple, but it took 60 years before the present author did it in 2002, and it seems nobody touched the issue in between. This chapter presents the analysis and deals with some minor details, also devising a dynamic process which shows fast convergence—a bit disappointing if one wants complex dynamics. In the stubs we also suggest putting the case in the two dimensions of geographical space, though here the integrals indeed become problematic, at least if one assumes an Euclidean distance metric. All becomes much simpler with a Manhattan metric, so this is what we suggest. The model is defined, but its analysis is left to the reader.

7.1 Introduction

7.1.1 Biographical

Harold Hotelling (1895–1973) was an American mathematical statistician. In this main field he was credited with the T-square distribution. He also made several ingenious contributions to mathematical economics, though so few that a commemorative volume of his economics papers provide for a very slim volume. To this slimness also depends on his style of presentation, always succinct and to the point. He has been credited with a Law, a Lemma, and a Rule, though what is called Hotelling's Law is largely a misinterpretation of what we are going to discuss. Probably some minor researcher wanted to add authority to his own ideas.

He produced a classical contribution "*The Economics of Exhaustible Resources*" in 1931, and "*Stability in Competition*" dating from 1929 which we are now going to focus.

The present author also thinks his "*Mathematical Theory of Migration*", an MA thesis from 1921, is a gem. It was republished in Environment and Planning through the care of Sir Alan Wilson, though it seems that only two economists in almost a Century, ever commented the paper. It was not even included in the commemorative volume mentioned. It models Malthusian growth according to Verhulst as a logistic process, combined with diffusion in space, from densely to sparsely populated areas. The problem is that it employs the Laplacian operator. As economists never cared to learn vector analysis the whole thing is likely to have remained a mystery to most.

Exactly that model was reinvented in ecology by Skellam 30 years later, and then discussed in thousands of publications. For this, Skellam was baptized "father of mathematical ecology".

It may seem that Hotelling's original contribution might be of interest for economists today when we have to deal with worldwide migration flows, but not so.

7.1.2 *Some History*

The enigmatic Hotelling model of duopoly on a fixed line interval appeared in 1929. It was the first and seemingly only well structured case of Bertrand oligopoly, as heterogeneity of a physically homogenous good was obtained through spatial separation of suppliers and accrued transportation costs.

Unfortunately, Hotelling only analyzed the case where demand was totally inelastic, which resulted in the paradox that both competitors would crowd in the centre. Yet, Hotelling himself in his verbal discussion stated that the paradox would disappear if demand were elastic—there would remain a tendency to gravitate towards the centre, rather than locating at the socially optimal quartiles, though without the extreme crowding.

It is a bit surprising that Hotelling did not himself follow this track. Perhaps he too was under the spell of paradox, like following authors, and so wanted to emphasize this particular case. There exists an immense literature in his aftermath. Unfortunately, citation databases do not extend as far back as to 1929, so it is difficult to give an exact estimate.[1]

Analytically, the crowding version of the model is problematic, and even inconsistent. With location in the same point, the space Hotelling introduced slips out again, and we are back to the problems that Bertrand pointed out. Inconsistency concerns consumer behaviour. They choose the lowest delivered price, which creates the market areas, but yet each customer buys one unit of commodity, no matter what its price.

Lerner and Singer in 1937, provided the first proof that the crowding paradox would evaporate with elastic demand. If one only assumed the consumers to have a reservation price then the crowding would not occur. The meaning of this is that if delivered price is higher than the reservation price then the consumers would buy nothing, otherwise just one unit as in Hotelling's model. The long paper is still most enjoyable through its ingenious use of graphic argument.

Very soon after, Smithies (1941a,b) in two articles put up the problem with a linear downsloping demand function, which seems to correspond to what Hotelling had in mind for the general case. However, Smithies claimed that the integrals were too complicated to evaluate, which is a mystery as they are quite simple.

So it was left to the present author, 60 years later, in 2001 to carry out the formal analysis of Smithies's case (Puu 2001, 2002). Dependent maximum demand price a,

[1]No doubt the paradox created its popularity. Scientists replaced distance by just "similarity" in some vague sense for competing products, or even for political ideologies, all with doubtful measurability. Vague analogies deprive such a good scientific model of its qualification as science. Further, to escape some consequences of paradox, the unrealistic and most contrived idea of quadratic transportation costs was launched by some mathematicians at CORE in Louvin la Neuve and, surprisingly, gained much popularity.

transport cost rate k, and unit production cost c, combined to one single composite parameter $\kappa = \frac{a-c}{k}$, three outcomes are possible:

1. Disjoint monopolies.
2. Genuine duopoly.
3. Cutting out monopoly.

The first case, occurs when transportation costs are high and the competitors can locate so widely apart that demand goes to zero before the potential territories reach each other. The third case is as ill structured as the Bertrand price cutting always was, and it still needs a truly convincing layout. The second case is genuine duopoly, well structured, and worthy of closer study. Note that with elastic demand crowding in the centre only occurs under extreme conditions, so it can be ignored if we want to focus more well structured cases.

Professor Laura Gardini and the present author studied the dynamics of this genuine duopoly case in 2002. It turned out to be very simple; just a contraction to equilibrium, with locations separated in space, not quite at the quartiles, but not crowding in the middle either—quite as conjectured in Hotelling's original article.[2]

7.1.3 Digression About Space

There are several issues we have to consider in connection with space; essentially four; (1) dimension, (2) boundedness, (3) topological properties, and (4) the metric properties of space.

Dimension

The natural dimension of economic/geographical space is 2D. However, almost all models either start from one dimension, or, more rarely, after outlining the problem in 2D, simplify to 1D—from surface to line. Sometimes we find the intermediate case of imposed circular symmetry, but that is deceptive, as any such model is still one dimensional, just swept out by a rotating radius vector.

[2]However, this analysis assumed that the competitors shared the space available as a duopoly in a common boundary point, and that *at the other ends each market extended to the boundary points of Hotelling's fixed interval.*

In a communication to the present author Dr. Helge Sanner, then at the University of Potsdam, pointed out that the last may not be true—the competitors might also end the markets where local demand dropped to zero. Unfortunately the present author has not been able to locate any publication by Dr. Sanner to cite on this.

Further thought, however, indicates that this would never happen. If one endpoint only extends to the point where demand vanishes, the competitor in question would always profit from moving its location until an endpoint of the fixed interval is reached. This is because more spatial symmetry of the market shape always increases profits.

It goes without saying that a simplification from plane to line makes one miss very essential issues. For instance, on a line a flow of traded merchandise from one point to another *must* pass any intermediate point, whereas in 2D all points can be avoided by a detour if it is inconvenient to pass due to congestion or anything else.

The case is quite similar to Abbot's "*Flatland*" (1884) where we encounter all the surprises an amoeba would experience in a three dimensional world, just like us in 4D time-space.

Boundedness

As for the second issue, boundedness, it seems that all interesting phenomena, even Hotelling's simplified 1D model, only occur in *bounded space*. An infinite line has no centre, and therefore no crowding in the centre. Neither has the perimeter of a circle, nor the surface of a sphere. Hotelling speaks of location along an intercontinental railway to make sense of the line, though it is essential that the railway has finite length and fixed endpoints.

A related issue is that, in continuous space, models using partial differential equations (PDE), always involve boundary conditions without which the PDE is incomplete and can produce completely different solutions.

Global Topology

Apart from the question of bounded space we have to consider the topological nature of the space we are dealing with. It seems that in our geographical living space only two topological genera are meaningful—the plane and the spherical surface (if we consider our entire globe).

This should be no serious issue, though the present author has seen economic applications of the "game of life" on a grid of cells where there is contamination among contiguous cells as to their state—on/off. It seems that the model of local action could easily be adapted to economic phenomena. However, the interesting phenomena such as wandering and oscillating clusters only arise in the global topological setting of a torus, which is totally meaningless in the geographical setting. It may be tempting to appropriate intriguing discoveries from interesting models, but if we do not care about realism it remains just sloppy analogy.

Distance Metric

Finally, we should recall that metric always is an integral part of space. In the physical sciences the Euclidean distance is the norm, and any deviation arises from some exploration in queer geometries. In geographical space, non-Euclidean distance metrics, such as the Manhattan, or the ring-radial are the rule if we judge assumptions by realism. The same is true whenever we combine modes of transport, such as air and surface.

As long as we keep to 1D modelling, there is no metric difference between the Euclidean (circular distance loci) and the Manhattan (square distance loci), but as soon as we want to enter the full 2D of geographical space, the difference becomes important.

It should be emphasized that metrics different from the Euclidean may not only increase realism in models, but they can also be simpler to use in computation. We mentioned Smithies's claim that with elastic demand the linear demand integrals in Hotelling's 1D model were too complex to evaluate, though, as we will see, they are quite straightforward.

However, using Euclidean distance for 2D integration of demand in a market area, or even just the market area, can be quite messy. This was already known to Wilhelm Launhardt in 1885. From "*Mathematische Begründung der Volkswirtschaftslehre*" comes the famous graphical construction with Launhardt "funnels". See Fig. 7.1.

To give a taste of the difficulties involved consider the Euclidean metric where

$$d = \sqrt{(x_2 - x_1)^2 + (y_2 - y_1)^2} \tag{7.1}$$

is the distance between points (x_1, y_1) and (x_2, y_2) in the plane. Suppose these points represent the locations of two firms. Further, assume the competitors charge mill prices p_1, p_2 respectively and that transportation costs k € per unit of distance.

Fig. 7.1 Launhardt funnels to determine market areas for three firms. Over the location of each firm the vertical stem indicates its charged mill price. The conic surface above indicates the accrued transportation costs. Where these cones intersect, local delivered prices break even and there is a market boundary, which is projected down on the geographical 2D surface. The shape of a cone assumes an Euclidean metric. Notably, Launhardt explored a more general case where transportation rates are different, say for instance for close substitutes such as coal and oil. If so the firm supplying coal (higher transport cost) will have a local enclosed market island, immersed in a sea in which the oil provider dominates. As there are enough difficulties we skip this complication

Then for a consumer located at some point (x, y) the delivered price offers are

$$p_1 + k\sqrt{(x_1 - x)^2 + (y_1 - y)^2} \tag{7.2}$$

and

$$p_2 + k\sqrt{(x_2 - x)^2 + (y_2 - y)^2} \tag{7.3}$$

respectively.

To obtain a polynomial equation for the boundary curve we have to equate these two expressions, and the resulting equation has to be squared twice to get rid of the roots which results in a quartic. Obviously, evaluating double integrals bounded by quartic equations is bound to be quite messy. It can be done with the computer, but such numerical output is not useful in a model—we need simpler explicit expressions.

Consumers buy from the firm which provides the commodity at the lowest local delivered price, mill price (the stem) in Fig. 7.1 plus transportation cost (the funnel widening with distance). Local prices from two suppliers break even where the funnels intersect, and the vertical projections of the intersection curves show the boundaries of market areas in geographical space.

The conical shape depends on the assumption of an Euclidean metric. As mentioned, we also have different possible metrics for a region in geometrical space. The most well known is the Manhattan metric with, say roads in the South-North and West-East directions. As travel is straight in these directions only, the distance loci become tilted squares rather than circles. Of course, in reality there is a finite density of roads, but still the points of equal distance (and transportation cost) lie on these tilted squares. We get dense loci if we assume infinite density of roads, which, once we approach realism by relating metrics to infrastructure, may sound absurd. Though, this is what we always do when using the Euclidean metric. It represents possibilities to move along straight lines in any direction between any pair of locations in space, something that only birds, but not even aeroplanes can do.

The distance between points (x_1, y_1) and (x_2, y_2) in the Manhattan metric is

$$d = |x_2 - x_1| + |y_2 - y_1| \tag{7.4}$$

i.e., just the sum of the horizontal and vertical differences of the location coordinates. Market boundaries are defined by

$$p_1 + k\,|x_1 - x| + k\,|y_1 - y| = p_2 + k\,|x_2 - x| + k\,|y_2 - y| \tag{7.5}$$

Depending on whether $x_1 \lesseqgtr x_2$ and $y_1 \lesseqgtr y_2$ we get the simplification

$$\pm x \pm y = \frac{p_2 - p_1}{2k} + \frac{x_1 + x_2}{2} + \frac{y_1 + y_2}{2} \tag{7.6}$$

The r.h.s., is a constant dependent on mill prices and locations of the firms. The second and third term represent the horizontal and vertical midpoints between the locations of the suppliers. Hence, if the mill prices are equal, the boundary curve passes through this midpoint, if not, the point is dragged in the direction of the cheapest supplier. The l.h.s. shows that the boundary is composed by straight lines with a slope of $\pm 45°$. Compared to the quartic for the Euclidean case, integrations may be expected to by much simpler.

In Fig. 7.2, we see the Launhardt funnels in the context of a Manhattan metric.

This long digression was, of course, due to a wish to generalize the Hotelling case to 2D, which has never been properly done. Not even the case to be considered as such a generalization has been defined: Is it three firms in an equilateral triangle? Or, is it four firms in a square?[3,4]

It is now time to return to the Hotelling case in 1D, and formulate the entire model under elastic demand.

[3]We have by no means exhausted the possible metrics. Instead of roads in two directions at right angles as basis for the metric, we could have three directions intersecting at 60° angles. In such case we deal with the distance metric

$$d = |u| + |v| + |w|$$

where

$$u = (x_2 - x_1) + \sqrt{3}\,(y_2 - y_1)$$
$$v = (x_2 - x_1) - \sqrt{3}\,(y_2 - y_1)$$
$$w = 2\,(x_2 - x_1)$$

The distance loci are now regular hexagons. We could consider a case of three (or six) competitors locating in such a hexagon.

Also triangular distance loci are a possibility, provided we keep to the case of roads in three directions, but assume that in each direction there is only one lane for traffic in only one direction. Try to work out the corresponding metric!

As the market area shapes conform to the metric of space, recall that there exist only three regular tessellations for the 2D plane, by triangles, squares, and hexagons. Hence the metrics discussed in this note are the natural candidates (in addition to the Manhattan) for attempting a generalization of the Hotelling problem.

[4]There are also other metrics. One is the Minkowski metric

$$d = \left(|x_2 - x_2|^\rho + |y_2 - y_1|^\rho\right)^{\frac{1}{\rho}}$$

of which we already encountered two cases, the Euclidean, $\rho = 2$, and the Manhattan $\rho = 1$. The other cases are interesting, though not useful in computation and for obtaining simple models.

It is noteworthy that we already discussed this as the formula for the CES production function, which may or may not be modelled after the much older Minkowski metric.

Fig. 7.2 Launhardt funnels for the case of a Manhattan distance metric. Note that the market boundaries become parallel straight lines, so calculations of market areas and demand may become much easier than in the Euclidean case

7.2 The Model

7.2.1 Assumptions and Notation

Consider a line interval $[-1, 1]$ in which two competitors take locations x_1, x_2. They charge mill prices p_1, p_2. Given a transport cost rate k, local price at x is $p_i + k|x - x_i|$ depending on from which supplier $i = 1, 2$ the commodity is bought.[5] The boundary point is

$$x = \frac{x_1 + x_2}{2} \mp \frac{p_1 - p_2}{2k} \tag{7.7}$$

(sign depending on whether $x_1 < x_2$ or $x_1 > x_2$).

The boundary point would be halfway between the locations of the competing firms if their mill prices were equal. If not, the point is dragged in the direction of the firm charging the higher mill price, thereby decreasing its market share.

However, it is possible that demand goes to zero before this boundary point is reached. When this happens the firms can have disjoint non-competing monopoly areas. The following discussion follows rather closely a recent conference presentation in Tokyo, Puu (2016).

[5]Mill pricing, where consumers pay for full transportation costs is our case. However, it is by no means the only possibility. As the competitors are monopolists in their market areas, they can also themselves provide for delivery and apply price discrimination, provided they do not charge more for transport than its actual cost. It is well known that with linear demand perfect discrimination implies charging for exactly half the transportation cost.

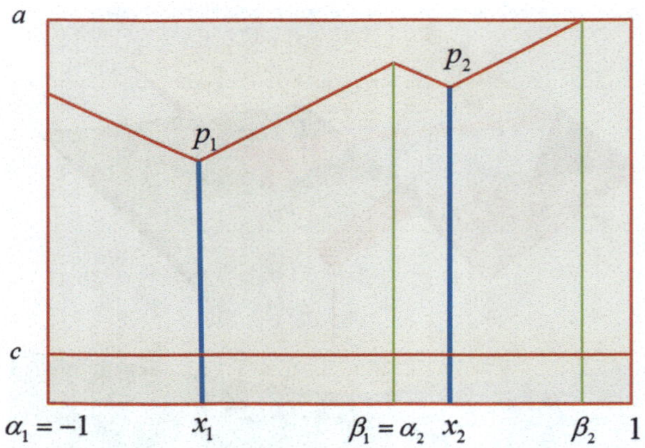

Fig. 7.3 Picture of the duopoly market case. For variety we displayed the market area of the left firm as extending to the interval boundary, whereas that of the right firm falls short of it, going only to the point where local demand vanishes due to a too high delivered price

Assume the linear demand function[6]

$$q_i = \max\left(a - \left(p_i + k\left|x - x_i\right|\right), 0\right)$$

Accordingly, demand for each competitor goes to zero when

$$x = x_i \pm \frac{a - p_i}{k} \tag{7.8}$$

We could call the expression $\frac{a-p_i}{k}$ market radius. Note that this applies both when we have two disjoint monopolies, and when one firm cuts the other out to establish one single monopoly. Market radius decreases with higher mill price and with higher transportation cost. High transportation cost would therefore favour establishing disjoint monopolies.

From (7.7) we get two candidates for market area endpoints, and (7.8) adds further four. Taking in account the fixed endpoints of the entire interval, chosen as ∓ 1, we are dealing with eight potential endpoints for market areas.

The geometry of the format is shown in Fig. 7.3.

[6]A more general formula for demand is $q_i = \max\left(a - b\left(p_i + k\left|x - x_i\right|\right), 0\right)$, but experience from work with linear models shows that parameter b has no independent influence, so we simplify by putting $b = 1$.

7.2.2 *Total Demand and Profits*

Let us so introduce symbols for these market area endpoints: α_i, β_i. Given these, we can calculate total demand

$$Q_i = \int_{\alpha_i}^{\beta_i} q_i dx = \int_{\alpha_i}^{\beta_i} a - (p_i + k\,|x - x_i|)\, dx \tag{7.9}$$

Despite Smithies's doubts, integration is simple:

$$Q_i = (a - p_i)\left(\beta_i - \alpha_i\right) - \frac{k}{2}\left((x_i - \alpha_i)^2 + \left(\beta_i - x_i\right)^2\right) \tag{7.10}$$

Profits then are

$$V_i = (p_i - c)\, Q_i \tag{7.11}$$

where c denotes constant unit production cost—for simplicity taken constant and equal for both competitors.[7]

Note in particular that the price minus cost factor in (7.11) does not depend on location. Hence, whenever we look for an optimal location, we can just maximize (7.10).

7.2.3 *Market Area Endpoints*

From our previous digression, it is obvious that the left endpoint of a market area α_i can take on the following values:

$$\alpha_i = \begin{cases} -1 \\ x_i - \frac{a-p_i}{k} \\ \frac{x_i+x_j}{2} + \frac{p_i-p_j}{2k} \end{cases} \tag{7.12}$$

and the right endpoint β_i

$$\beta_i = \begin{cases} \frac{x_i+x_j}{2} - \frac{p_i-p_j}{2k} \\ x_i + \frac{a-p_i}{k} \\ 1 \end{cases} \tag{7.13}$$

[7]It should be noted that Hotelling assumed *production cost to be zero*, like many other oligopoly theorists, such as Cournot, and von Stackelberg. Equal, but nonzero costs do not complicate things notably, and provide a more reasonable first approximation.

Obviously we can combine each left market endpoint α_i with each right endpoint β_i, except the case $\alpha_i = \frac{x_i+x_j}{2} + \frac{p_i-p_j}{2k}, \beta_i = \frac{x_i+x_j}{2} - \frac{p_i-p_j}{2k}$ as in a shared market the left endpoint of the firm to the right cannot be combined with the right endpoint of the firm to the left.

This leaves eight cases[8]:

$$[\alpha_i, \beta_i] = \begin{cases} \left[-1, \frac{x_i+x_j}{2} - \frac{p_i-p_j}{2k}\right], & \mathbf{1} \\ \left[-1, x_i + \frac{a-p_i}{k}\right], & \mathbf{2} \\ [-1, 1], & \mathbf{3} \\ \left[x_i - \frac{a-p_i}{k}, \frac{x_i+x_j}{2} - \frac{p_i-p_j}{2k}\right], & \mathbf{4} \\ \left[x_i - \frac{a-p_i}{k}, x_i + \frac{a-p_i}{k}\right], & \mathbf{5} \\ \left[x_i - \frac{a-p_i}{k}, 1\right], & \mathbf{6} \\ \left[\frac{x_i+x_j}{2} + \frac{p_i-p_j}{2k}, x_i + \frac{a-p_i}{k}\right], & \mathbf{7} \\ \left[\frac{x_i+x_j}{2} + \frac{p_i-p_j}{2k}, 1\right], & \mathbf{8} \end{cases} \tag{7.14}$$

This list exhausts all possibilities for total market demand, as only endpoints matter in (7.10). In most cases the evaluation is quite simple. A graphic summary of the regions is displayed in Fig. 7.7.

Also recall that only total demand matters for the choice of optimal location.

7.2.4 *Interpretation of the Cases*

Shared Duopoly Market

The duopoly cases are listed as **1**, **4**, **7**, and **8**. See Fig. 7.3. In all they are four, because given the location of one competitor, the other may choose to locate to the left or to right of it. Further, its market at the outer end can extend to the respective interval endpoint, or stop at the point where local demand vanishes. These are the most interesting cases of the Hotelling model, as they deal with genuine duopoly. As mentioned, cases **1** and **8** were investigated in Puu and Gardini, and it was about the cases **4** and **7** that the author was alerted by Dr Sanner.

Monopolies

Cases **2**, **5**, and **6** represent monopoly. We can either have two coexistent disjoint monopolies, Fig. 7.4, not sharing any common boundary, or one competitor cutting

[8]Actually they represent even more cases as a generic firm i could locate to left or right. Further a monopoly case can represent disjoint monopolies, or cutting out the competitor.

out the other, Fig. 7.5. Case **3** too is a cutting out monopoly that covers the entire fixed interval. This case is what has been focused in the literature.

7.2.5 *Demand and Optimal Location*

We can now use the endpoints listed in (7.14) to calculate total demand for the firms according to (7.10). Recall that the choice of location in all cases can be done through maximizing demand, as the multiplicative factor $(p_i - c)$ in the expression for profits (7.11) does not include location.

After finding optimal locations that maximize demand, we can substitute it out, and obtain expressions for profits that only depend on mill price. So, let us take the list (7.14) case by case.

Cases 1 and 8 (Duopoly)

These true oligopoly situations are the only tough cases. It was these that, however, were investigated in Puu and Gardini and shown to be simple contractions in the dynamic adjustment process.

In cases **1** and **8** the market of firm 1 extends from the left endpoint $\alpha_1 = -1$ of the interval to a common market boundary point $\beta_1 = \frac{x_1+x_2}{2} - \frac{p_2-p_1}{2k}$. Substituting in (7.10), we get

$$Q_1 = (a - p_1)\left(1 + \beta_1\right) - \frac{k}{2}\left((1 + x_1)^2 + \left(\beta_1 - x_1\right)^2\right) \tag{7.15}$$

Maximizing through putting the derivative $\frac{\partial Q_1}{\partial x_1} = 0$ we can solve for a new optimum location

$$x_1' = \frac{x_2 - 4}{5} + \frac{2a - 3p_1 + p_2}{5k} \tag{7.16}$$

Note that, as a consequence of choosing x_1' according to (7.16), the right market boundary point is changed to

$$\beta_1' = \frac{3x_2 - 2}{5} + \frac{a - 4p_1 + 3p_2}{5k} \tag{7.17}$$

This consequence of location choice is perceived by the firm, so, along with x_1' from (7.16), β_1' from (7.17), must be substituted back in its estimate of demand.

Defining

$$\lambda_1 = p_2 + k\,(1 + x_2) \tag{7.18}$$

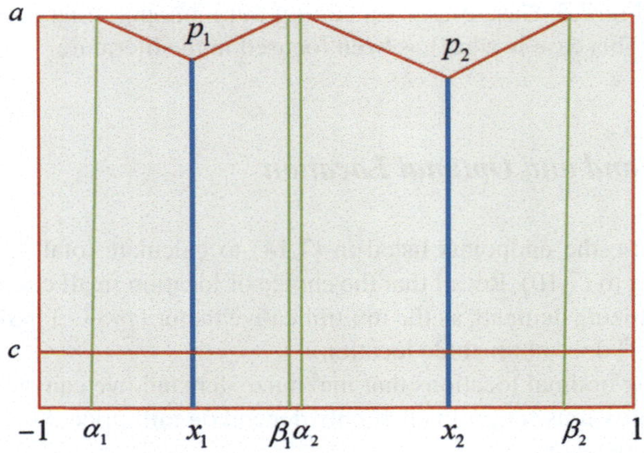

Fig. 7.4 A case of disjoint monopolies that do not compete on the fixed interval. Such a case is most likely to occur when transportation costs are relatively high, so that the monopoly areas within which local demand remains positive can be accommodated within the fixed interval. The exact locations of the firms for this case are not determinate

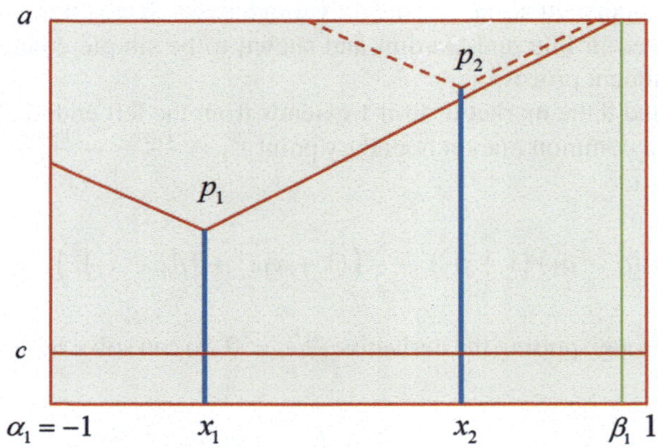

Fig. 7.5 Cutting out monopoly. Through lowering mill price the left firm makes delivered price at the location of the right firm lower than the mill price of the latter, so taking the entire market. This will, of course, be contested by the cut out firm in further moves

the resulting total demand becomes

$$Q_1' = \frac{1}{10k}\left(6\,(a-p_1)^2 - 4\,(a-p_1)\,(a-\lambda_1) - (a-\lambda_1)^2\right) \tag{7.19}$$

Likewise for firm 2, whose market extends from $\alpha_2 = \frac{x_1+x_2}{2} - \frac{p_2-p_1}{2k}$ to $\beta_2 = 1$, substitution in (7.10) yields

$$Q_2 = (a-p_1)\,(1-\alpha_2) - \frac{k}{2}\left((\alpha_2 - x_2)^2 + (1-x_2)^2\right) \tag{7.20}$$

Putting $\frac{\partial Q_2}{\partial x_2} = 0$ and solving, we obtain

$$x_2' = \frac{x_1+4}{5} - \frac{2a - 3p_2 + p_1}{5k} \tag{7.21}$$

Again this changes the left market boundary point to

$$\alpha_2' = \frac{3x_1+2}{5} - \frac{a - 4p_2 + 3p_2}{5k} \tag{7.22}$$

which, again, along with substitution of the new location, must be substituted back to get a total demand estimate. Defining

$$\lambda_2 = p_2 + k\,(1-x_2) \tag{7.23}$$

we get

$$Q_2' = \frac{1}{10k}\left(6\,(a-p_2)^2 - 4\,(a-p_2)\,(a-\lambda_2) - (a-\lambda_2)^2\right) \tag{7.24}$$

Note two things; (i) the slight change of sign in (7.18) as compared to (7.23), and that, given this, (7.19) and (7.24) have exactly the same form.

Also, note that (ii) λ_1, λ_2 contain all information about the competitor's location and mill price. About these any firm can only have expectations, and, as usual, we will take the simplest possibility—that each firm believes the other to retain their previous moves, even if they do not in a dynamic process.

This makes total demand, given optimal location choice, quadratic in its own mill price for each firm.

Then, multiplication by $(p_i - c)$ makes profit a cubic in mill price. So, once we proceed to optimize with respect to mill price, first order conditions result in a quadratic to solve—though it is obvious which of its two solution roots provides profit maximum. This analysis is carried out below.[9]

Cases 4 and 7 (Duopoly)

We can now pass to the cases where the firms still share a common market boundary point, but the other market endpoint falls short of the interval limit because local demand goes to zero before it is reached. I will call it the Sanner Case.

Again there are two cases, depending on which side of the prelocated competitor the firm chooses.

Surprisingly these cases are considerably simpler than the previous. Taking case **4**, the endpoints of the market are $\alpha_1 = x_1 - \frac{a-p_1}{k}$, the point where local demand vanishes, and $\beta_1 = \frac{x_1+x_2}{2} - \frac{p_1-p_2}{2k}$, where the delivered prices of the firms break even.

Substituting in (7.8) and putting $\frac{\partial Q_1}{\partial x_1} = 0$, we get the optimal location

$$x_1' = x_2 - \frac{2a - p_1 - p_2}{k} \tag{7.25}$$

which substituted back in (7.8) yields

$$Q_1' = \frac{1}{k}(a - p_1)^2 \tag{7.26}$$

This is a formula we will encounter repeatedly in the sequel.

In case **7**, location is to the right of the competitor, between points $\alpha_2 = \frac{x_1+x_2}{2} + \frac{p_1-p_2}{2k}$ and, $\beta_2 = x_2 + \frac{a-p_2}{k}$, again substituting in (7.10), and putting $\frac{\partial Q_2}{\partial x_2} = 0$ we get optimal location

$$x_2' = x_1 + \frac{2a - p_i - p_j}{\kappa} \tag{7.27}$$

Substituting back in (7.10) we again find the same formula as (7.26)

$$Q_2' = \frac{1}{k}(a - p_2)^2 \tag{7.28}$$

Now there is a simple argument why the Sanner Case will never be chosen in duopoly: Given any mill price, total demand will increase with the degree of symmetry of the market, which is obtained through approaching the interval limit

[9] We chose to denote the firm to the left 1, and the one to the right 2. This works in each single step, but it is fully possible, and shows up in numerical experiment, that each firm can move to the other side of its competitor, and so we must keep track of the numbering.

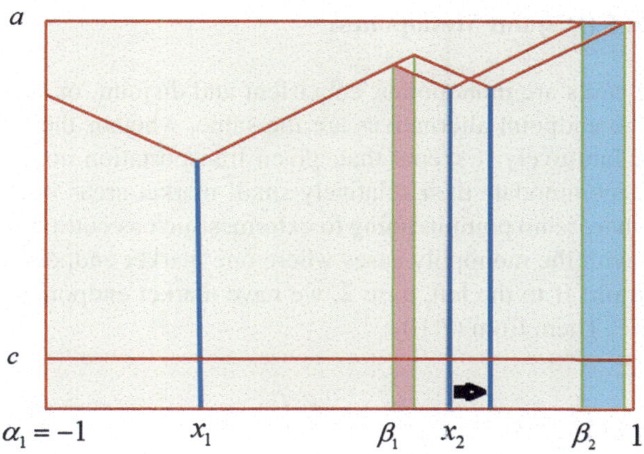

Fig. 7.6 Gain (blue) and loss (red) strips of total sales when the firm to the right takes a step towards the right boundary of the interval. Note that it does not quite stretch out the market for the firm to the right to this right boundary—so, there remains a little further possible gain in sales

∓ 1 on the side of the firm. Then, either such move results in monopoly if there is space enough, or in duopoly with a larger market demand. Further, the competitor has no reason to retaliate at any stage because by such move the firm decreases competition in the boundary point.

Digression on the Sanner Case

We can make this clear by a simple graphic argument. Take the case illustrated in Fig. 7.6, where the market for the firm to the right does not stretch out to the right interval endpoint. Then assume the firm changes location as indicated by the arrow.

The blue vertical strip to the right indicates the gain of total sales, the red one to the left, the loss to the competitor. As we see, the loss area is much smaller than the gain area.

When the firm changes location, the gain area on the right has its boundary pushed exactly the same distance as the change of location. However, the left market area is moved much less, due to the sloping delivered price line of the competitor. This is, of course, due to transportation cost, but the exact level for it does not matter—the left loss strip remains narrower than the right gain strip. Further, in terms of area (total sales), it is bounded above by the delivered price line, whereas the gain strip to the right goes all way up to maximum chargeable price.[10]

[10]Of course, a change of location will be followed by a change of mill price. But this is a further issue, location can be chosen so as to maximize sales alone.

Cases 2, 6, and 5 (Disjoint Monopolies)

The remaining cases are monopolies, coexistent and disjoint, or single cutting out monopolies. The endpoint alternatives are the same, whether there is coexistence or cutting out. Intuitively it seems that, given transportation costs are high, and the firms can accommodate their relatively small market areas in the preassigned segment, then there is no point in going to extremes such as cutting out.

Let us start with the monopoly cases where one market endpoint extends to the fixed interval limit. If to the left, case **2**, we have market endpoints $\alpha_1 = -1$ and $\beta_1 = x_1 + \frac{a-p_1}{k}$. Then, from (7.10)

$$Q_1 = (a - p_1)\left(1 + x_1 + \frac{a - p_1}{k}\right) - \frac{k}{2}\left((1 + x_1)^2 + \left(\frac{a - p_1}{k}\right)^2\right) \tag{7.29}$$

Putting $\frac{\partial Q_i}{\partial x_1} = 0$,

$$x_1' = \frac{a - p_1}{k} - 1 \tag{7.30}$$

which substituted back yields

$$Q_1 = \frac{1}{k}(a - p_1)^2 \tag{7.31}$$

i.e., again the same formula as (7.26).

For case **6**, the firm locates on the right, with market limits $\alpha_2 = x_2 - \frac{a-p_2}{k}$ and $\beta_2 = 1$,

$$Q_2 = (a - p_2)\left(1 - x_2 + \frac{a - p_2}{k}\right) - \frac{k}{2}\left((1 - x_2)^2 + \left(\frac{a - p_2}{k}\right)^2\right) \tag{7.32}$$

The optimal location is

$$x_2' = 1 - \frac{a - p_2}{k} \tag{7.33}$$

Upon substitution back it again yields

$$Q_1 = \frac{1}{k}(a - p_1)^2 \tag{7.34}$$

quite as (7.31).

Finally, let us so consider case **5**, a monopoly where the market does not extend to any of the interval endpoints. It extends from $\alpha_i = x_i - \frac{a-p_i}{k}$ to $\beta_i = x_i + \frac{a-p_i}{k}$, i.e., as far as where demand vanishes at either end. Now there is no need to distinguish between locations to left or right.

Substituting for the endpoints in (7.10) we obtain

$$Q_i = \frac{1}{k}(a - p_i)^2 \tag{7.35}$$

i.e., again the same formula as in (7.31), (7.33), (7.26) and (7.28).[11]

Case 3 (Cutting Out Monopoly)

Remains just one case **3**, a monopoly extending over the whole interval $[-1, 1]$. This can only be a cutting out monopoly. From (7.10) we immediately obtain

$$Q_i = 2(a - p_i) - \frac{k}{2}\left((1 + x_i)^2 + (1 - x_i)^2\right) \tag{7.36}$$

Optimizing with respect to location we find[12]

$$x_i' = 0 \tag{7.37}$$

and substituting back the simple expression

$$Q_i' = 2(a - p_i) - k \tag{7.38}$$

7.2.6 *Profit Maximization and Mill Price*

We now have all the total demand expressions for all possible market area endpoint $[\alpha_i, \beta_i]$ combinations. Likewise, we derived optimal location choices, unless they were indeterminate. We can hence pass to considering profits as functions of mill prices, and optimize with respect to these.

---

[11]Notably, this time we did not optimize with respect to location to get the formula. Demand, and therefore profit as well, is *independent* of location, and location itself *indeterminate*. Therefore, we have the disadvantage of not getting a definite location choice for the map we want to formulate. To solve this problem, consider that we deal with coexistent disjoint monopolies, whose maximum profits do not depend on location. We can choose either -1 or 1. This actually means merging the case with cases **2** or **6**.

[12]Note that this case can only occur when transportation cost is very low. As the entire interval $[-1, 1]$ must be covered, market radius $\frac{a-p}{k}$ must exceed unity, i.e., half the interval.

Cases 1, 8

Start with the toughest cases, where the interval is split in duopoly, and the market areas stretch out to the interval endpoints at their far ends. Demand for cases **1** and **8** was given in (7.19) and (7.28), which had the same form given the different definitions (7.18) and (7.23)

$$Q_i' = \frac{1}{10k}\left(6\,(a-p_i)^2 - 4\,(a-p_i)\,(a-\lambda_i) - (a-\lambda_i)^2\right) \tag{7.39}$$

Substituting in the profit expression $G_i' = (p_i - c)Q_i'$ it becomes a cubic in p_i,

$$V_i' = (p_i - c)Q_i' \tag{7.40}$$

The optimality condition $\frac{\partial V_i}{\partial p_i} = 0$ then yields a quadratic in p_i:

$$18p_i^2 - 4(4a + 3c + 2\lambda_i)p_i + a^2 + 2a(4c + 3\lambda_i) + \lambda_i\,(4c - \lambda_i) = 0 \tag{7.41}$$

The quadratic, as usual, has two roots, of which the smaller provides profit maximum:

$$p_i' = \frac{4a + 3c + 2\lambda_i}{9} - \frac{\sqrt{36(a-c)^2 - 24\,(a-c)\left(a-\lambda_j\right) + 34\left(a-\lambda_j\right)^2}}{18} \tag{7.42}$$

Locations could now be obtained from (7.16) and (7.21) through substitution of p_i, but if we write them down, they just look messy. In numerical work it is easy to let the computer do the job.

Cases 2, 4, 5, 6, 7

As we saw, demand in all these cases is given by one single formula, $Q_i = \frac{1}{k}\,(a - p_i)^2$. These are two cases of duopoly with one end of the market limited by vanishing local demand (Sanner cases, **4** and **7**), and three cases of monopoly (**2**, **5**, and **6**), either coexistent or cutting out.

In all these cases profits are

$$V_i = \frac{1}{k}\,(p_i - c)\,(a - p_i)^2 \tag{7.43}$$

Thus the optimum condition

$$\frac{\partial G_i}{\partial p_i} = \frac{1}{k}\,(a - p_i)^2\,(a + 2c - 3p_i) = 0 \tag{7.44}$$

has two zeros $p_i = a$, and

$$p_i = \frac{1}{3}a + \frac{2}{3}c \tag{7.45}$$

The first is a minimum resulting in zero profit as mill price is equal to maximum chargeable price, which results in vanishing demand everywhere, even at the location of the mill. Expression (7.45) is a well known formula for such linear models—to charge a weighted average of maximum price and unit production cost, with weight one third for the first and two thirds for the latter. See Beckmann (1968).

Also maximum profit using (7.45) is now easy to calculate

$$V_i = \frac{4}{27k}(a-c)^3 \tag{7.46}$$

Locations

Once we know optimal price for these five cases, we can calculate optimal locations from (7.25), (7.27), (7.30), (7.33), and (7.37).[13]

Thus

$$\mathbf{2,6}\ x_i = \pm\frac{2}{3k}(a-c) \mp 1 \tag{7.47}$$

$$\mathbf{4,7}\ x_1 = \pm\frac{1}{k}(p_2 + kx_2) \mp \frac{1}{3k}(5a-2c) \tag{7.48}$$

$$\mathbf{5}\ x_i = \pm\left(\frac{2}{3k}(a-c) - 1\right) \tag{7.49}$$

Case 3

From (7.38) we get profit

$$V_i = (p_i - c)(2(a - p_i) - k) \tag{7.50}$$

[13]Note that though price according to (7.45) is in the right interval between c and a, it is by no means certain that the location according to these formulas is reasonable—depending on parameters an "optimal" location may even fall outside the admissible interval $[-1, 1]$. We will therefore have to run a check of relevance with respect to region for each alternative once we proceed to formulating the map. Note further that in the monopoly cases (except for cutting out) the location like optimal price is independent of the competitor's moves, whereas in the shared market cases **4** and **6** the best reply depends on the expected move by the competitor.

which, optimized with respect to price, yields

$$p_i = \frac{1}{2}a + \frac{1}{2}c - \frac{1}{4}k \tag{7.51}$$

Substituted back, we obtain

$$V_i = \frac{1}{2}\left(a - c - \frac{k}{2}\right)^2 \tag{7.52}$$

This is a cutting out case with location chosen in the midpoint of the interval as we see from (7.37). The case corresponds to the Hotelling paradox, and will no doubt be contested by the competitor. Recall what was said about the conditionality of this case on a low transportation cost, and also that there are other instances of cutout behaviour.

7.3 Summary for the Cases

To interpret Fig. 7.7, note that, taking x_2, p_2 as given, we can use the same type of diagram as in all the pictures above as phase space for x_1', p_1', the dash, as usual, denoting the next move considered. Several features should be noted.

The empty areas in the rectangle, below c, and in the little triangle on top of the competitor's mill price, are such which firm 1 never considers—a mill price lower than unit cost yields no profit, and neither does a combination of location and mill price where the firm cuts itself out.

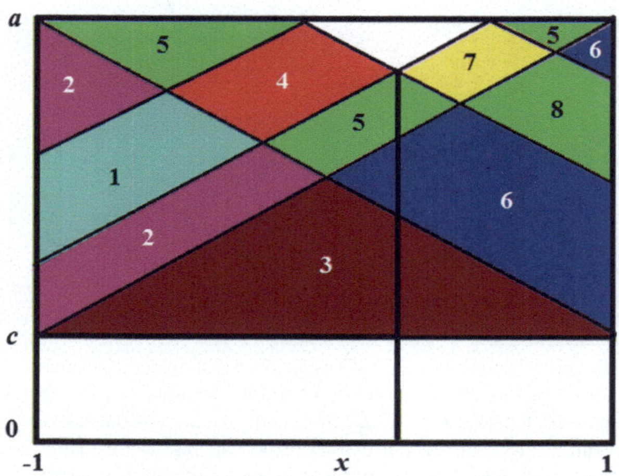

Fig. 7.7 Possible moves for x_2', p_2' with regions relevant for the eight endpoint combinations **(1)**, ... **(8)**. when x_1, p_1 is fixed

The structure seems quite complicated, but we see that the regions are separated by parallel straight lines, three downsloping and three upsloping, whose formulas are easy to state.

Define $r_i = p_i + kx_i, \quad s_i = p_i - kx_i$. Then the downsloping lines have formulas

$$\begin{aligned} r_1 &= 2a - r_2 \\ r_1 &= r_2 \\ r_1 &= a - k \end{aligned} \tag{7.53}$$

and the upsloping

$$\begin{aligned} s_1 &= 2a - s_2 \\ s_1 &= s_2 \\ s_1 &= a - k \end{aligned} \tag{7.54}$$

Further study, using our formulas for optimal location and mill pricing choice, show that the *optimal* x_1', p_1' points in the plane can only be located on stretches of these sloping lines, which, of course, simplifies analysis a lot. We, however, do not present the argument here as it leads into further detail.

Notably, there are three areas labelled with **5**, those on top representing local monopolies, left respectively right of the competitor. The area mid in the picture is a single cutting out monopoly. Also regions **2** and **6** are in duplicate, local monopoly extending to the interval boundary on top, and a corresponding cutting out case lower down. In all we see 12 coloured regions, but three of them are further split by the vertical line at the competitor's location, so if we count locations to the left and right of the competitor, we arrive at 15 different cases.

After computing profits, we choose the reaction which returns the highest profit.

Note that for dynamics the results are quite sufficient for writing a simple computer program, though the map is too complicated with all its pieces to formulate any elegant closed form iteration formula.

7.4 Equilibria

All numerical experiments show that there are no periodic or more complicated orbits for the map, provided we ignore the cutting out cases, which remain too ill-structured to be included. However, there exist two different kinds of equilibrium states; genuine duopoly with a common boundary, and disjoint non-competing monopoles.

7.4.1 Duopoly

The most interesting, though messy, case is oligopoly when the firms share the total market between them with some competition at the common boundary, cases **1** and **8**.

Locations

As the firms are identical (facing the same demand and having the same production and transport costs), it is a reasonable conjecture that in equilibrium they also charge the same mill prices, i.e. $p = p_1 = p_2$. Substituting in (7.16) and (7.21), putting $x_1' = x_1$, $x_2' = x_2$, now taking the equations as a simultaneous system, and solving, we get

$$x_1 = \frac{1}{3}\frac{a-p}{k} - \frac{2}{3} \tag{7.55}$$

$$x_2 = \frac{2}{3} - \frac{1}{3}\frac{a-p}{k} \tag{7.56}$$

Adding (7.55) and (7.56)

$$x_1 + x_2 = 0 \tag{7.57}$$

Obviously the firms locate symmetrically around the centre, which seems plausible by intuition.

Let us pay attention to two particular cases; (1) if $\frac{a-p}{k} = \frac{1}{2}$ then $x_1 = -\frac{1}{2}$, $x_2 = \frac{1}{2}$, and (2) if $\frac{a-p}{k} = 2$ then $x_1 = x_2 = 0$. In the first case the firms locate socially optimally at the quartiles, each at the centre of its half-interval, in the second they crowd both in the centre, Hotelling's main case.

Hence we see that according to (7.55)–(7.56) both these outcomes are possible under duopoly action, as are any cases between these. The locations depend on one single compound expression, $\frac{a-p}{k}$, the difference of maximum chargeable price and actual mill price, divided by the transportation rate.

Demand

Note that with equal mill prices, and symmetric locations from (7.17) and (7.22) we have $\beta_1 = \alpha_2 = 0$, so the market areas are $\alpha_1 = -1, \beta_1 = 0$ and $\alpha_2 = 0, \beta_2 = 1$.

Further, from (7.18) and (7.23)

$$\lambda = \lambda_1 = \lambda_2 = \frac{4p - a + 5k}{3} \tag{7.58}$$

Substituting this in (7.19) and (7.24), along with $p_1 = p_2 = p$, we obtain

$$Q_1 = Q_2 = \frac{6\,(a-p)^2 - 4\,(a-p)\,(a-\lambda) - (a-\lambda)^2}{10k} \tag{7.59}$$

Price and Profit

From (7.11), then

$$V_i = (p-c)\,Q_i \tag{7.60}$$

We already obtained the optimal duopoly prices in (7.42) above. As we deal with equilibrium, we can remove the dash, delete the index on costs and mill prices, and substitute for λ_1, λ_2 from (7.58).

By this (7.42) becomes quite messy. Furthermore, substitution for λ_1, λ_2 introduces mill prices under the root sign. Therefore, we have to square out the root again and solve for mill price anew. As the resulting equation is quadratic in mill price, we again have two roots and have to choose the relevant one, obtaining:

$$p = \frac{2a + 3c + 8k}{5} - \frac{3\sqrt{4(a-c)^2 - 8\,(a-c)\,k + 34k^2}}{10} \tag{7.61}$$

We can restate this as

$$\frac{a-p}{k} = \frac{3}{5}\kappa - \frac{8}{5} + \frac{3}{10}\sqrt{4\kappa^2 - 8\kappa + 34} \tag{7.62}$$

where

$$\kappa = \frac{a-c}{k} \tag{7.63}$$

Equations (7.55)–(7.56) let us determine x_1, x_2 once we know $\frac{a-p}{k}$. This ratio was already designated as an important compound, but it depends on mill price, which is a decision variable. The ratio $\kappa = \frac{a-c}{k}$ has the same form, but is more fundamental as it only contains parameters.

We can now use (7.61) and (7.58) in (7.59). Using the convenient (7.63), total demands become

$$Q_1 = Q_2 = \frac{k}{50}\left((22 - 2\kappa)\sqrt{4\kappa^2 - 8\kappa + 34} + 2\kappa^2 - 24\kappa + 67\right) \tag{7.64}$$

Further, using (7.60),

$$V_1 = V_2 = \frac{k^2}{250}((2\kappa^2 - 44\kappa + 377)\sqrt{4\kappa^2 - 8\kappa + 34} + 4\kappa^3 - 92\kappa^2 + 482\kappa - 2194) \tag{7.65}$$

—not so elegant, but closed form expressions for total demands and profits anyhow.

As mentioned, this case was analyzed in Puu and Gardini (2002). Any dynamic process defined by Eqs. (7.16), (7.21), (7.18), (7.23), and (7.42) is just a contraction. In Fig. 7.8 we illustrate how fast this global approach to equilibrium goes.

In only five steps the system has converged so close to equilibrium that further iterations can no longer be distinguished. Not seen in the picture, but showing up when we print the locations, is that in the first move the firms swap left/right positions. Firm 2 starts so far to the right that it finds it better to move to the left of firm 1.

7.4.2 Monopolies

As for the monopoly cases, we already derived optimal prices, profits, and locations, unless they were indeterminate. Given parameter κ is high, for instance resulting from a high transportation rate, only narrow market areas delimited by positive local demand can be established. These can be easily accommodated in the fixed interval, and there is no point in crowding to a duopoly. In these cases location

Fig. 7.8 Illustration for the contraction to equilibrium duopoly. In the foreground the first move is displayed after a random initial. In the background six moves later, no divergence from the final state is detectable. As we see the contraction to equilibrium price and location is extremely fast

Fig. 7.9 Iteration to disjoint monopolies using the duopoly map. Again the contraction in but few iterations leads to the final state. In this case we designed a combination of parameters such that the disjoint monopolies would exactly cover the interval. If they cover less, due, for instance, to a higher transport cost, then the final locations would be indeterminate, though the contraction map still works for finding the positions where the market areas just touch in the middle

is indeterminate, because its exact choice does not influence demand or profit, provided the fixed interval endpoints and the other competitor's monopoly area are not touched.

Should the competitor be too centrally located, the firm will temporarily have to choose a duopoly, but the competitor will then find it more profitable to establish a monopoly area more to its side. This is confirmed by numerics. See Fig. 7.9.

We can even use the iterative system designed for duopoly, as it also works for disjoint monopolies. Several interesting things can be noted in this case. The convergence is again rather fast. Now, we take a starting position where the firms are initially located very close to the centre. Further, the initial prices are so chosen that firm 1 is actually cut out by firm 2. Yet *both* firms in the next move jump wide apart—even so wide that they hit the interval boundaries. This, we already noted, is inoptimal, so after that the firms start converging to their final positions, though definite positions as defined by the duopoly process. As a matter of fact, once they are disjoint monopolies, their exact final positions are indeterminate.

7.5 Further Issues

In a first analysis one can assume, as Hotelling originally, that both competitors are free to choose location *and* mill price in each move.

7.5.1 *Different Time Scales*

However, we could also consider a combination of short run dynamics where only prices change but location is fixed, and a long run dynamics where both location and price are variable. In such a case it is natural to connect location change with capital formation. A firm is unlikely to change location as long as it has capital invested in some plant. But capital wears out with time, and when it is time for renewal, then also location can be changed.

The present author has in mind capital with constant rate of decay and "sudden death", combined with a Leontief type of production function which produces capacity limits.

7.5.2 *Different Pricing Policies*

Above we mentioned that price discrimination was not considered. Any competitors with local monopoly market areas could consider price discrimination, arranging for transportation and charging less than actual transportation cost. It is even possible to combine discrimination in an inner area with mill pricing at the fringe towards the competitor. See Beckmann (1968). Also note that uniform pricing is a case of discrimination, though utterly unsuited for the Hotelling case as one does not get any definite market boundary points.

References

Beckmann MJ (1968) Location theory. Random House, Inc., New York
Hotelling H (1929) Stability in competition. Econ J 39:41–57
Lerner AP, Singer HW (1937) Some notes on duopoly and spatial competition. J Polit Econ 45:145–186
Puu T (2001) Bertrand oligopoly revisited. Discret Dyn Nat Soc 6:1–10
Puu T (2002) Hotelling's 'ice cream dealers' with elastic demand. Ann Reg Sci 36:1–17
Puu T (2016) Hotelling duopoly revisited. In: Matsumoto A, Szidaraovsky F, Asada T (eds) Essays in economic dynamics, theory, simulation analysis, and methodological study. Springer Science + Business Media Singapore Pte Ltd., Berlin, pp 35–56. ISBN 978-981-10-1520
Puu T, Gardini L (2002) Hotelling type duopoly and oligopoly. In: Puu T, Sushko I (eds) Oligopoly dynamics: tools and models. Springer, Berlin, pp 265–310. ISBN 3-540-43186-1
Smithies A (1941a) Optimum location in spatial competition. J Polit Econ 49:423–439
Smithies A (1941b) Monopolistic price policy in a spatial market. Econometrica 9:63–73

Chapter 8
Disequilibrium Trade and Pricing of Durable Commodities

Abstract The model proposed focuses on trade in disequilibrium with stocks or assets, i.e., durable commodities. Durability implies that the goods have a second hand value and that trade is possible through redistribution of fixed totals of commodity stocks. The first connotation in connection with stocks refers to financial items; bonds, shares, and money, items that only function as stores of value. However, we take a broader perspective, to commodities that also yield services that enter the utility functions. The best example is perhaps items of real estate, apartments and suburb villas, which we can take as the two commodities. Demand naturally shifts over time due to normal demographic change. Retired people move to central apartments, whereas families with children need the space of suburban villas. If real estate prices drift up, everybody becomes nominally more wealthy, so demographic change can cause substantial value drifts without any speculation being involved. Real estate preserves value more than most other durables, yet for most consumers they are just habitations. They also have the property that their total changes very little over time, so that they can be taken as fixed totals redistributed predominantly in disequilibrium. In this chapter we take the case of two agents and two commodities, so that we can use the familiar Edgeworth "box" with the two traders represented in opposite corners. Each agent has a simple Cobb-Douglas set of indifference curves, of opposite concavity. The curve of tangency points between the two sets, the traditional "cont(r)act" curve, provides an infinite set of possible equilibrium states. New is the setting of durable commodities, property distributions represented through points in the box that need not be on the "cont(r)act" curve. An external tâtonnement agent is assumed to devise a price search mechanism based on excess demand. The relative price announced determines the slope of the budget line, but new in the context is that it rotates through the **actual** property distribution point inside the box. Even if equilibrium cannot be attained at the price announced, the traders may yet find a deal profitable for both. If so, in general the amounts they wish to exchange are different, so it is assumed that the trader wanting to exchange less limits the deal. Then an exchange will take place, though it as a rule will not reach equilibrium, and there will remain some excess supply/demand after exchange. One can object that real estate items are "quantized", but what is not in reality? It may also be the case that both want to exchange the same commodity for the other, and then no trade can take place. The dynamic system devised then sticks

T. Puu, *Disequilibrium Economics*, https://doi.org/10.1007/978-3-319-74415-5_8

to such a disequilibrium point, while price goes on oscillating, always overshooting equilibrium price. It all depends on the adjustment step, if it is small, an equilibrium will be approached though very slowly, if it is big, approach is faster but price may overshoot equilibrium, and the exchange system stick to a disequilibrium for ever. It is difficult for the external agent to find out these things as the outcome may be very high period oscillation or even chaotic. Other conclusions are that the system may approach different equilibrium or disequilibrium states, depending on the initial point and adjustment step, and that exchange traces are irreversible, i.e., return to a previous price will **not** make the system return to a previous state. Unlike standard economics, unique demand and supply functions exist as little as do unique equilibria.

8.1 Introduction

Recent instabilities in the economy have been recognized as, if not entirely due to, so at least triggered by market bubbles, the housing market being a typical case. The models used for the explanation now invariably are of the "heterogenous agent" types, based on interactions between "fundamentalists" and "chartists", i.e. speculators. The latter try to identify value trends and coming turning points in order to make maximum short run profits, the former believe in asset prices returning to their long fundamental values. This tradition seems to originate in a paper by Day and Huang from 1990 with the catchy title "Bulls, bears, and market sheep".

No doubt a habitable apartment or house primarily has some value in terms of service utility for the owners, but in view of how much actual market values change over time, one may ask whether any of these can be designated as fundamental. It also seems little convincing to ascribe value variations to speculation, because relatively few people speculate in their own habitations. Apartments and houses always change hands due to demographic facts; young families need more space when they get children, retired people need less space, etc. If market values are high, then everyone is nominally richer. Through exchange market values can drift up (or down) without anybody actually speculating in value changes. One can hence find basic reasons for the volatility of asset prices due only to the simple fact that the commodities are traded and redistributed but remain on the market. Modelling speculative behaviour is not even necessary, which is a benefit. As Lord Keynes so eloquently pointed out, speculation, even though, unfortunately, an annoying fact of real life, is a phenomenon so irrational that it is almost hopeless to model.

Economists are used to analyze market phenomena in terms of supply and demand functions and their equilibrium intersections. This works fine with flows, such as food stuff; these are consumed during one period and then influence nothing more. Stocks or assets, i.e., any kind of durables, on the contrary, remain on the market.

Through trade their ownership is just redistributed. This simple fact has overlooked consequences: When budget lines rotate due to changed prices, the consumers find new optimal positions in terms of their indifference maps. If some want to sell a commodity, and others want to buy it, then trade takes place, and ownership is redistributed. However, once trade *has* taken place, budget lines will rotate through *new* points. If then, after a change, prices return to original values, the agents' optimal choices will not return to the original positions, which even no longer are on the new budget lines. Hence, demand and supply reactions to changing prices are *irreversible* once trade has taken place.[1]

Demand and supply functions become dependent on the actual trading history, and they cannot be uniquely defined; neither can their equilibrium intersections.

In what follows this distinguishing fact for assets, that they transfer wealth in kind from one time period to another will be focused, and its consequences will be studied within the simplest possible setting of a pure trade model with two agents and two commodities as once formalized by Edgeworth in his ingenious box model.

Conditions for trade in *dis*equilibrium now have to be specified. This issue is never addressed in economics which focuses on equilibria. Even when at an announced market price one agent wants to sell and the other to buy a certain commodity it would be an unlikely coincidence if the intentions of the agents match so perfectly as to result in equilibrium. We have to state what exchange would take place when one agent wants to buy more than the other wants to sell or the other way around. In the case of two agents, it is quite obvious. It may be advantageous for both to move even part of the way towards the desired optima, but there is no way to force any agent to move beyond his optimum position. Thus we state that the new point after exchange is set by the agent who wants to move less.[2]

It is also possible that at an announced price both agents want to buy/sell the same commodity. In this case, obviously, no trade can take place, so the system remains in the point where it was before.

To close the system, we must state how the prices are set. The case with two agents may seem to be a natural case for bilateral monopoly, but we intend this stylized formulation only as a first stepping stone for a system with more agents and more commodities. Hence, one can assume a Walras type of tâtonnement process where prices announced depend on excess demand/supply as suggested by Samuelson. In discrete time it is appropriate to choose a semilogarithmic dependence of price on excess demand/supply, in order to avoid prices becoming negative.

It may also be interesting to shift analysis from Walras's tâtonnement to Marshall's supply and demand price. Then a lot quantity, not an announced price, is the given datum for the agents. This mimics real auctions where the supply price

[1]The present author raised this issue around 1965, and was fortunate to discuss it with several outstanding economists, such as Robert Clower, Leif Johansen, and Tord Palander (now all deceased), but he never published anything except a short note in Puu 1965. By that time path dependence and hysteresis were not yet on the agenda, so there was no obvious method to apply.

[2]Once we have more agents this becomes tricky; one must state which agents can move closer to the intended optima and which have to remain more dissatisfied with large remaining excess demand/supply. See the respective research stub.

is the reservation price. Once more traders are added we also get possibilities of bidding. This would also take care of any preoccupation with quantized objects in trade.

8.2 Assumptions

8.2.1 Notation

Denote the commodity quantities for the first agent (x, y). As it makes no harm in a pure exchange model, normalize the totals of both commodities available on the market to unity. Hence, the corresponding quantities for the second agent are denoted $(1 - x, 1 - y)$. The Edgeworth box thus becomes a unit square. Any actual distribution of assets is denoted in upper case, (X, Y) for the first agent, $(1 - X, 1 - Y)$ for the second.

We only deal with the relative price, normalizing the price of the first commodity to unity, which just becomes the numéraire. The price of the second commodity is denoted p, which hence is a relative price.

8.2.2 Budget Constraints

The budget constraint for the first agent reads

$$x + py = X + pY \tag{8.1}$$

and for the second agent,

$$(1 - x) + p\,(1 - y) = (1 - X) + p\,(1 - Y) \tag{8.2}$$

However the latter is identical with (8.1), which we see if we subtract the expression $(1 + p)$ from both sides of (8.2).

This almost is all notation we need. Let us just call the optimal points to which the agents would like to move respectively (x_1, y_1) and (x_2, y_2), and we are finished. These points obviously have to lie on the budget line (8.1) or (8.2), quite as the actual wealth distribution point (X, Y). The points (x_1, y_1) and (x_2, y_2) touch an indifference curve each from the preference map of either agent. All this is illustrated in Fig. 8.1. Obviously the system cannot move from (X, Y) to both optimal points. Below we will discuss to which of them the new wealth distribution point $\left(X', Y'\right)$ actually moves through trade.

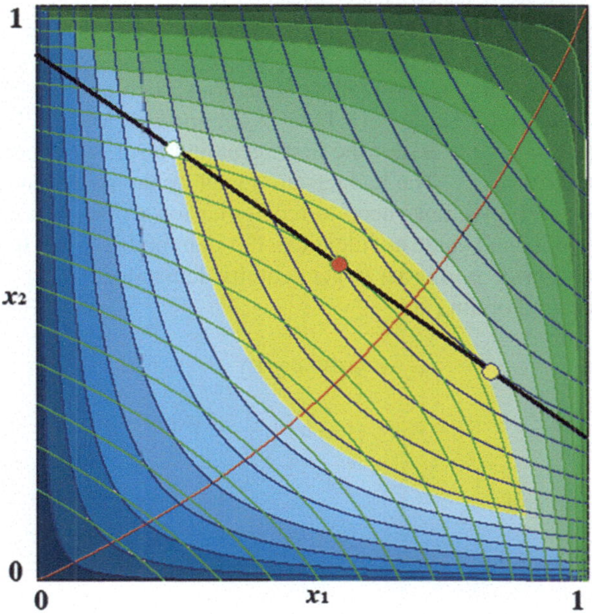

Fig. 8.1 Indifference maps for the two competitors $U = x^{\alpha} y^{1-\alpha}$ (blue), $V = (1-x)^{\beta} (1-y)^{1-\beta}$ (green), and the budget line $x + py = X + pY$. Initial point (X, Y) (in white), and the points (x_1, y_1) (yellow) and (x_2, y_2) (red), to which the agents would like to move, are displayed. Further on display is the curve for equilibria, along which the indifference curves touch, and the yellow lens shaped area, in which both agents are better off than in the initial point. Parameter values are: $\alpha = 0.6$, $\beta = 0.4$, $X = 0.25$, $Y = 0.75$, $p = 1.5$. Given these, from (8.5) $x_1 = 0.825$, $y_1 = 0.3\overline{66}$, and from (8.6) $x_2 = 0.55$, $y_2 = 0.55$. As we see, the yellow (x_1, y_1) is farther away from the initial white point (X, Y) than is the red point (x_2, y_2), so the second agent would like to move farther along the budget line than would the first. However, agent 1 has no means to force agent 2 to move that far, and he yet profits from moving to the red point which is inside the "lemon" of points better than the initial for both. Hence, the second trader limits exchange, yet exchange takes place though equilibrium is not attained

8.2.3 Utility Functions

As for utility, take a Cobb-Douglas form for the utility function. Then, for the first agent the utility function is

$$U = x^{\alpha} y^{1-\alpha} \tag{8.3}$$

and for the second

$$V = (1-x)^{\beta} (1-y)^{1-\beta} \tag{8.4}$$

Obviously, the Cobb-Douglas exponents must be in the interval $0 < \alpha, \beta < 1$.

8.2.4 Individual Optima

The results of maximizing Cobb-Douglas functions, such as (8.3)–(8.4) subject to linear budget constraints, such as (8.1) or (8.2), are well known: Fixed shares of the budget, equal to α, $(1-\alpha)$ for the first agent and β, $(1-\beta)$ for the second are spent on commodities x, y. The budgets are given by the right hand sides of (8.1) and (8.2) respectively. We only need to recall that for x price is unity, so, for that commodity, quantity and value are identical. For commodity y one has to divide the budget share with price p in order to get quantity demanded.

Hence the desired optima are

$$x_1 = \alpha\,(X + pY) \tag{8.5}$$
$$y_1 = (1-\alpha)\left(\frac{X}{p} + Y\right)$$

for the first agent, and

$$(1 - x_2) = \beta\,((1 - X) + p\,(1 - Y)) \tag{8.6}$$
$$(1 - y_2) = (1-\beta)\left(\frac{1 - X}{p} + 1 - Y\right)$$

for the second. Note that, in addition to the exponents of the utility functions, (8.5) and (8.6) depend on relative price p and on the actual asset distribution X, Y, or $(1 - X)\,, (1 - Y)$.

As a rule (x_1, y_1) and (x_2, y_2) are different; from each other and from the actual asset distribution point (X, Y), as we see in Fig. 8.1. Only in one situation are they equal, i.e. on the equilibrium curve which is also illustrated in Fig. 8.1. It can be obtained in different ways. We can put $x_1 = x_2$ and $y_1 = y_2$ in (8.5), (8.6) and eliminate p; or we can calculate the locus of points where the indifference curves, one from each utility map, touch (through calculating and equating the derivatives). The latter is the usual way used in international trade theory where the equilibrium curve is called "contract curve". Whatever the procedure, the formula reads

$$Y = \frac{(1-\alpha)\,\beta X}{\alpha\,(1-\beta) + (\beta - \alpha)\,X} \tag{8.7}$$

which too is shown in Fig. 8.1. When $\alpha = \beta$ it is a straight line, the diagonal.

Note that the equilibrium relative price p in each point of (8.7) is well defined. It equals the slope of the touching indifference curves in that point. Also note that along the curve (8.7) there is a nondenumerable infinity of different equilibrium points.

8.2.5 Trade

Above it was stated that trade is limited to the *smallest* change that any agent wants to make, because the agent wanting to change more would then benefit from moving part of the way towards his optimum, whereas he has no means to force the other agent to move further than he wants. Referring to Fig. 8.1, we see that in the case illustrated this means moving from (X, Y) to $\left(X', Y'\right) = (x_2, y_2)$. Once this move has taken place, the budget line would pivot through this *new* point at any further change of the relative price p.

Supposing price changes from p to $p\prime$ and then back to p again, the original asset distribution point (X, Y) would *not* be retrieved, because after trading this point would not even be on the new budget line. For this reason one can neither speak of unique demand or supply functions in the case of assets, nor of unique equilibria. Any asymptotic equilibrium state depends on the intervening trading process. There are infinitely many equilibria, and any of these can be the asymptotic state depending on where the process starts and how the tâtonnement for pricing works. Even worse, as we will see, there are also infinitely many disequilibrium fixed points. This is the nucleus of the present argument.

The trade possibilities can be classified in six distinct categories, based on how the points (X, Y), (x_1, y_1), and $(x_2.y_2)$ are ordered, from left to right in the Edgeworth box. Actually, we can just use the x coordinate for this ordering.

Six Cases

(i) $x_1 \leq x_2 < X$: The first agent wants to sell $(X - x_1)$ but the second is only willing to buy $(X - x_2)$. There can be trade, but it is limited by the buyer's willingness. As a consequence one gets the new trade point $\left(X', Y'\right) = (x_2, y_2)$. There is left an excess supply amounting to $(x_2 - x_1)$, and a corresponding excess demand for the other commodity. Note for future use that for this case $(x_2 - x_1)(x_2 - X) \leq 0$.

(ii) $x_2 \leq x_1 < X$: The first agent wants to sell $(X - x_1)$ and the second wants to buy $(X - x_2)$, which is more. Trade is now limited by the seller's, the first agent's willingness. The new trade point becomes $\left(X', Y'\right) = (x_1, y_1)$. There is left an excess demand amounting to $(x_1 - x_2)$, and a corresponding excess supply of the other commodity. Note that for this case $(x_1 - x_2)(x_1 - X) \leq 0$.

(iii) $X < x_1 \leq x_2$: The first agent wants to buy $(x_1 - X)$ and the second wants to sell $(x_2 - X)$, which is more. Trade is limited by the buyer's, i.e., the first agent's offer. The new trade point becomes $\left(X', Y'\right) = (x_1, y_1)$. There is left an excess supply amounting to $(x_2 - x_1)$. For this case $(x_1 - x_2)(x_1 - X) \leq 0$.

(iv) $X < x_2 \leq x_1$: The first agent wants to buy $(x_1 - X)$ and the second wants to sell $(x_2 - X)$, which is less. Trade is limited by the seller's, i.e., the second agent's offer. The new trade point is $\left(X', Y'\right) = (x_2, y_2)$. There is left an excess demand amounting to $(x_1 - x_2)$. For this case $(x_2 - x_1)(x_2 - X) \leq 0$.

(v) $x_1 < X < x_2$: The first agent wants to sell $(X - x_1)$ and the second as well wants to sell $(x_2 - X)$. As both want to sell the same commodity, no trade is possible, so $\left(X', Y'\right) = (X, Y)$. There is left an excess supply amounting to $(x_2 - x_1)$. For this case $(X - x_1)(X - x_2) < 0$.

(vi) $x_2 < X < x_1$: The first agent wants to buy $(x_1 - X)$ and the second as well wants to buy $(X - x_2)$. As both want to buy the same commodity, no trade is possible. Again $\left(X', Y'\right) = (X, Y)$. There is left an excess demand amounting to $(x_1 - x_2)$. For this case too $(X - x_1)(X - x_2) < 0$.

The very simple argument is that if x_1 and x_2 are on either side of X, then trade is impossible (cases (v) and (vi)), because both agents want to buy/sell the same commodity. If x_1 and x_2 are on the same side of X, then there is one potential buyer and one seller, but the change is limited by the agent who wants to buy or sell least, agent 1 in cases (ii) and (iii), or agent 2 in cases (i) and (iv). The map thus boils down to three cases

$$\left(X', Y'\right) = \begin{cases} (x_1, y_1) \\ (x_2, y_2) \\ (X, Y) \end{cases}$$

The Trade Map

The map thus just reads

$$\left(X', Y'\right) = \begin{cases} (x_1, y_1) \text{ if } (x_1 - x_2)(x_1 - X) \le 0 \\ (x_2, y_2) \text{ if } (x_2 - x_1)(x_2 - X) \le 0 \\ (X, Y) \text{ if } (X - x_1)(X - x_2) < 0 \end{cases} \tag{8.8}$$

The application clauses exhaust all logical possibilities and are mutually exclusive as can easily be established. The two first rows of (8.8) represent trade corresponding to the limits set by agent 1 and 2 respectively, whereas the last row represents blocked trade because the agents want to buy or sell the same commodity. The weak inequality signs in the first two branches let us include the equilibria in the map.

8.2.6 Excess Demand

Let us restate the definitions for the desired optimal points to be used in (8.8), (x_1, y_1) and (x_2, y_2). They were given in (8.5) and (8.6), though it is nicer to solve for y_1 and y_2 in explicit form. Hence

$$\begin{pmatrix} x_1 \\ y_1 \end{pmatrix} = \begin{pmatrix} \alpha (X + pY) \\ (1 - \alpha)\left(\frac{X}{p} + Y\right) \end{pmatrix} \tag{8.9}$$

$$\begin{pmatrix} x_2 \\ y_2 \end{pmatrix} = \begin{pmatrix} 1 - \beta\left((1 - X) + p\,(1 - Y)\right) \\ 1 - (1 - \beta)\left(\frac{1-X}{p} + 1 - Y\right) \end{pmatrix} \tag{8.10}$$

This almost completes the iterative map; only one item is missing, the setting of relative price p.

8.2.7 *Tâtonnement*

As mentioned, the case of two traders may seem to set the stage for bilateral monopoly, but as the model is intended for generalization to more traders, it is better to use a Walrasian type of tâtonnement process where price is adjusted to excess demand.

As we deal with relative price p, we can choose either excess supply on the market for x or excess demand on the market for y to trigger price increases as there is a simple reciprocity between them in the model. We go for the first alternative. From the six cases (i)–(vi) listed above we find that excess supply, or if negative, excess demand for x always equals $x_2 - x_1$.

A linear price adjustment function could easily lead to negative prices which we want to avoid, so we choose the semilogarithmic,

$$p' = p \exp\left(\delta\,(x_2 - x_1)\right) \tag{8.11}$$

where δ denotes an adjustment step length. The choice of the map (8.11) has the advantage of symmetry with respect to the other relative price $1/p$, as the exponent then just changes sign.

8.3 The Model

The dynamic model we propose consists of (8.8) and (8.11), where (x_1, y_1) and (x_2, y_2) are as defined in (8.9)–(8.10). Despite its simple look, the model seems to be too complicated for further closed form analysis. We can, however, obtain much information through numerical experiment.

8.4 Numerical Analysis and Graphics

8.4.1 *The Phase Plane*

Trade Equilibria and Disequilibria

It is easy to run the map (8.8) and (8.11) with definitions (8.9)–(8.10) on the computer and display the results in a phase plane such as Fig. 8.1. One just

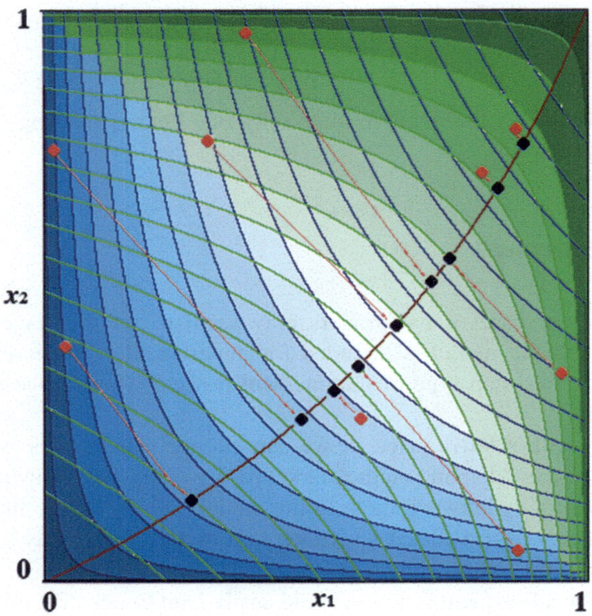

Fig. 8.2 Orbits for randomly generated initial points (X, Y) using the map 100,000 iterations. Parameters: $\alpha = 0.6$, $\beta = 0.4$, $\delta = 1$, where the moderate value of the adaptation coefficient is significant. Notably, all the orbits from initial red dots to final black dots converge to the equilibrium curve, though to *different* points. Accordingly there are an infinity (continuum) of equilibrium points. Initial relative price is set at $p = 1.5$, though p is adjusted in each iteration according to (8.11). In contrast to the next picture, for higher adaptation coefficient $\delta = 5$, the final black dots are on the contact curve. The low adaptation coefficient prevents overshooting of price adjustments. However, notably the final approach to equilibria is, as may be expected, slow as it takes up to 100,000 iterations. (The first steps are big, though, as we see from the red traces where the iteration stations are indicated by small dots.) It may be an interesting agenda to find out how many iterates it actually takes to arrive close to equilibrium (to end up just there, of course takes an infinite number)

needs to set the parameters, the exponents of the utility functions α, β, and the price adjustment step length δ, and further choose the initial values for the asset distribution (X, Y) as well as for relative price p. Figures 8.2 and 8.3 show the orbits generated from a set of randomly selected initial points (X, Y) in the unit square. As in Fig. 8.1 we keep the parameter values $\alpha = 0.6$, and $\beta = 0.4$. The initial relative price was set at $p = 1.5$ and the system was run in 100,000 iterations in each case. The iterates are indicated by small circles joined by line segments. Obviously, the first steps are giant, and the orbits converge fast on the final positions.

The difference between Figs. 8.2 and 8.3 is due to the step size, $\delta = 1$ in Fig. 8.2, $\delta = 5$ in Fig. 8.3. With the smaller step size all orbits converge to the equilibrium curve as displayed in Fig. 8.2, though to different points, thereby illustrating what was said about the dependence of equilibrium upon the dynamic adjustment process.

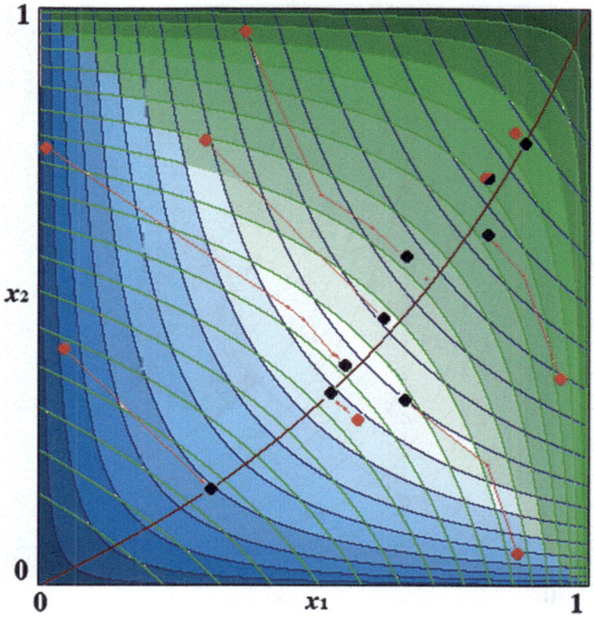

Fig. 8.3 Orbits of the same initial points (X, Y) as in the previous figure. Initial relative price is still $p = 1.5$ and the parameters: $\alpha = 0.6$, $\beta = 0.4$, but now the step adaptation coefficient is increased to $\delta = 5$. Again the orbits converge, though most of them to *disequilibrium* fixed points (it seems that only three end up at the curve of equilibria). The rest eventually stick to points where no further trade is possible because overshooting price fluctuations make the system oscillate between various excess supply/demand situations where both agents always want to sell/buy the same commodity. One orbit marked with a half red half black dot will be studied closer in the following pictures. In this case the system never moves at all from the initial point, which is both beginning and end of the orbit in quantity space. Note also that the adjustment traces this time sometimes seem to make sudden changes of direction. It would be interesting to know why

In Fig. 8.3, some orbits seem to again converge to the equilibrium curve, but some stop at a distance from it. Visually this stopping in disequilibrium fixed points also occurs in few steps; the large number of iterations suggests that the process indeed does not leave these final disequilibrium fixed points. Notably, for some initial points the process never leaves these points. From the discussion above we know what these disequilibria signify—cases where both agents want to buy/sell the same commodity so that no further trading is possible. The excess demand triggered tâtonnement process simply fails to reach an equilibrium point.

The process never makes one single move from one of the disequilibrium fixed points in Fig. 8.3, and we are going to study the price dynamics for this particular point in some detail.

To get some more information about disequilibrium fixed points, instead of just generating a few initial phase points, as in Fig. 8.3, in Fig. 8.4 we run the process from all initial phase points, packed as close as the resolution admits and mark just

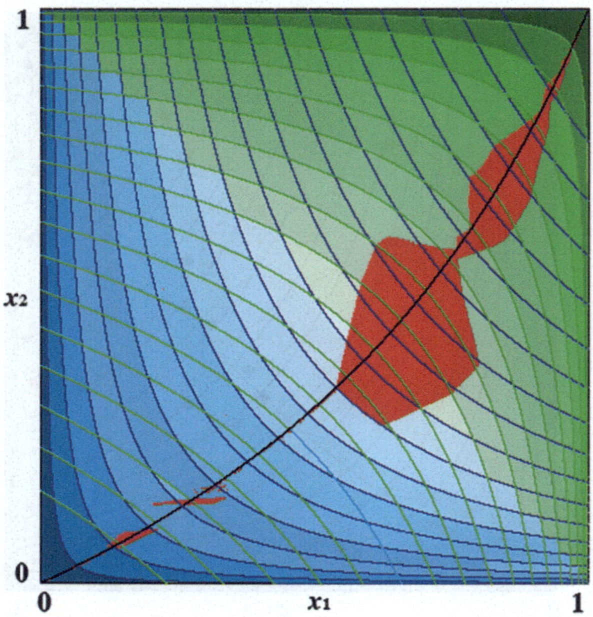

Fig. 8.4 Final fixed points of the orbits from *all* initial points (X, Y) in the square (or rather distinct pixels on the computer screen), for the same case as in the previous picture. The number of iterations from each initial point was reduced to 5000. Initial relative price was again $p = 1.5$. Parameters: $\alpha = 0.6$, $\beta = 0.4$, and $\delta = 5$ as in the previous picture. The *disequilibrium* fixed points agglomerate to the neighbourhood of the equilibrium curve, but occasionally swell out over considerable areas of the square. The previous picture indicated that the system with overshooting price adaptation only occasionally ended up at the equilibrium curve; now we take a more systematic view of the entire square

the final fixed points. As we see they cover curves and areas in the phase plane. To make the computation manageable the number of iterates for each orbit was reduced from 100,000 to 5000. The area of fixed points in Fig. 8.4 seems to accumulate around the equilibrium curve known from Figs. 8.1, 8.2, and 8.3, sometimes thin as a curve, sometimes swelling out to structures with nonzero area measure.

Price Oscillations

An interesting feature of the model is that the tâtonnement process produces continued price dynamics, periodic or aperiodic. This is illustrated in Figs. 8.5, 8.6, 8.7, and 8.8, all produced for the parameter combinations $\alpha = 0.6$, $\beta = 0.4$ quite as in Figs. 8.1, 8.2, 8.3, and 8.4. The initial price was set to $p = 1.5$ and the initial asset distribution $(X, Y) \approx (0.74, 0.55)$ as corresponding to the dot we spoke of in Fig. 8.3.

The parameter that takes on different values in this series of illustrations is the step size parameter; $\delta = 5$ in Fig. 8.5, $\delta = 5.1$ in Fig. 8.6, $\delta = 5.2$ in Fig. 8.7, and $\delta = 5.25$ in Fig. 8.8.

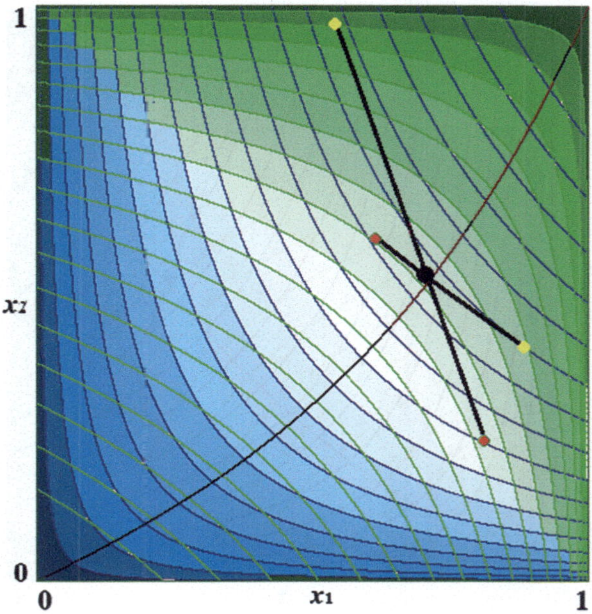

Fig. 8.5 A 2-period relative price oscillation from a disequilibrium fixed point (X, Y), with initial relative price $p = 1.5$. Parameters: $\alpha = 0.6$, $\beta = 0.4$, and $\delta = 5$. Shown are flipping budget line segments with endpoints (x_1, y_1) (yellow) and (x_2, y_2) (red). Note that they swap their positions relative to (X, Y). Hence in each new iteration excess supply is turned into excess demand and vice versa, which is due to the price overshoot. But as the yellow and red dots (the optima) always are on either side of the initial black point, the agents always want to sell/buy the same commodity, so no trade can ever occur. It may seem to involve an assumption of stupid rigidity on the part of the agents or the auctioneer. This may be in the case of a 2-period cycle, but any slight attempt to change adaptation may, as we will see, throw the system into something much more complicated that one can not easily adapt to

These pictures display the indifference maps and equilibrium curve in the phase plane; further the disequilibrium fixed point (X, Y) and a number of optimum points for the agents (x_1, y_1) and (x_2, y_2). These come in pairs and are joined by line segments, two in Fig. 8.5, four in Fig. 8.6, and six in Fig. 8.7. Note that all line segments pass the point (X, Y). These line segments are actually segments of the budget lines. In Fig. 8.8 the line segments are deleted and the end point pairs crowd dense along curves. These curves can be obtained in closed form through eliminating p in (8.5) and (8.6) respectively,

$$y_1 = \frac{(1-\alpha)\,Y x_1}{x_1 - \alpha X}, \tag{8.12}$$

$$y_2 = \frac{(1-\beta)\,(1-Y)\,x_2}{x_2 - \beta\,(1-X)} \tag{8.13}$$

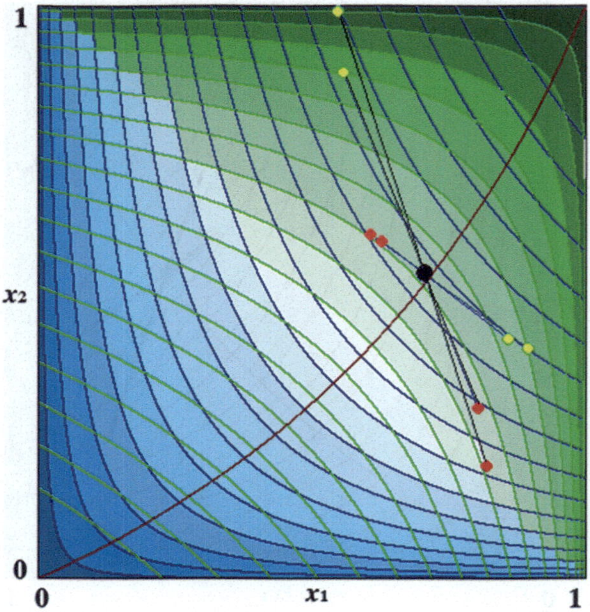

Fig. 8.6 The same case as the previous, but with a 4-period relative price oscillation when the adaptation coefficient increases to $\delta = 5.1$. Again the optimal points (yellow and red) switch relative positions in each iteration

The curves have been superposed on the numerically calculated trains of budget segment endpoints in order to show that this indeed is so.

The mechanism can be explained referring to Fig. 8.5. The relative price p oscillates between two different values and so the budget line flips between two different slopes. The endpoints (x_1, y_1) and (x_2, y_2), optima for the agents, are always on either side of the fixed point (X, Y), but they switch positions when the relative price oscillates; at one value both want to sell, at the other both want to buy the same commodity. Excess demand and excess supply alternate and the tâtonnement process for pricing overshoots adjustments. In Fig. 8.6 this 2-period oscillation has changed to a 4-period, and in Fig. 8.7 to a 6-period oscillation, and in Fig. 8.8 to something aperiodic. Again the budget line flips between the different positions, and the endpoints swap their positions so that there is always excess demand or supply of the same commodity.

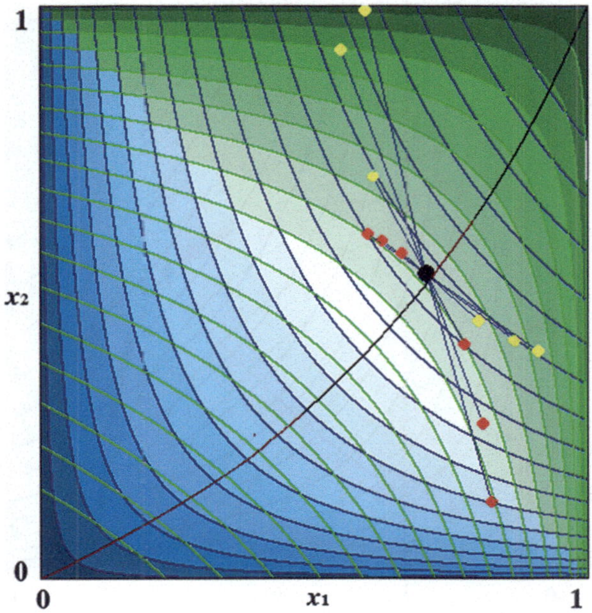

Fig. 8.7 Change to 6-period orbit when the adaptation coefficient increases to $\delta = 5.2$. Now, the agents, and tâtonnement auctioneer are starting to have difficulties, and can no longer be blamed for stupidity

8.4.2 Bifurcation Diagrams

Step Size Bifurcations

The relative price dynamics displayed in Figs. 8.5, 8.6, 8.7, and 8.8 can be summarized by the bifurcation diagram shown in Fig. 8.9. We now display p versus δ. The initial asset distribution point (X, Y) as well as the initial relative price p were kept to the fixed values used in Figs. 8.5, 8.6, 8.7, and 8.8, as were the parameters α, β. At each value of $\delta \in [4, 6]$ the system was run for 10,000 iterations. The first 9000 were trashed in order to get rid of transients, and the last 1000 were then plotted. If there is a fixed point then the same p will eventually be hit over and over. We just see a point, or, considering different adjacent δ producing fixed points, a line or curve. Once the fixed point bifurcates to a 2-period cycle we see the curve split in 2 branches, and so on in a cascade, eventually seeming to cover entire areas.

We see that the case of $\delta = 5$ shown in Fig. 8.5 fits into the 2-branch region, whereas at $\delta = 5.1$ shown in Fig. 8.6, there has been a further period doubling to 4. Then, after a stretch of (possibly chaotic) intervals, at $\delta = 6.2$ there are clearly 6 curve branches complying with Fig. 8.7. For $\delta = 5.25$ a dense vertical stretch is shown in accordance with Fig. 8.8.

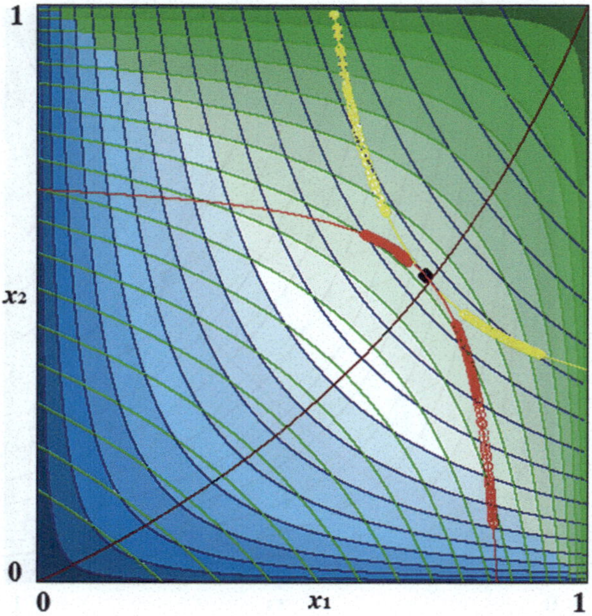

Fig. 8.8 When step size is increased to $\delta = 5.25$, the simple periodicity of relative price oscillation disappears. The points (x_1, y_1), (x_2, y_2) crowd densely on the curves (8.12) and (8.13), or on the edges of the box. Note that we could derive these curves $y_1 = \frac{(1-\alpha)Yx_1}{x_1-\alpha X}$ and $y_2 = \frac{(1-\beta)(1-Y)x_2}{x_2-\beta(1-X)}$ analytically through elimination of the relative price in the optimum conditions for the traders

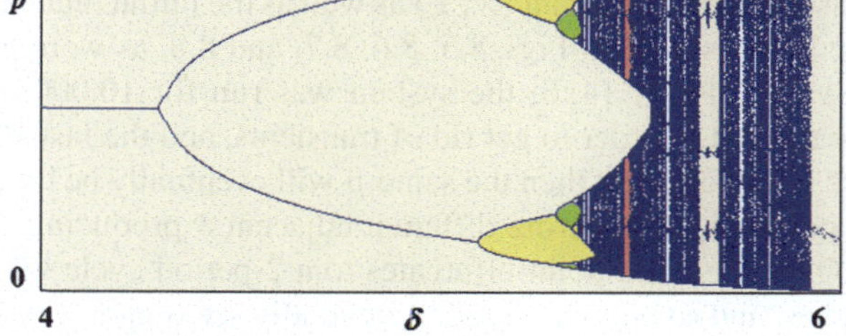

Fig. 8.9 Bifurcation diagram showing eventual relative price oscillations, periodic and aperiodic, as dependent on the step size parameter δ. As usual, we iterated the dynamical system, now with all the adaptation step sizes in the interesting interval, 10,000 times, trashed the first 9000, and plotted p for the remaining iterations. If the outcome is a fixed point, then the same value of p will be hit 1000 times (in the resolution of the screen), if the outcome is a 6-period cycle, the point plots will be distributed among the six different points. Looking at this process as a movie when the parameter increases, we see the bifurcations when the curves split. Obviously the predominant scenario is multi-periodic or chaotic

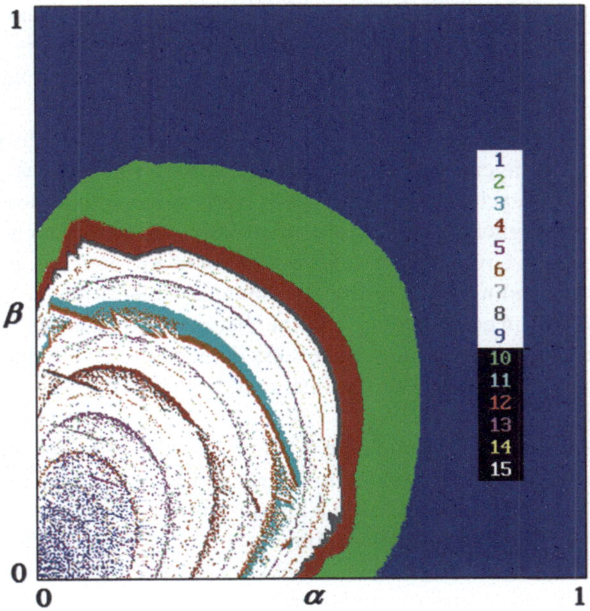

Fig. 8.10 This picture shows the bifurcation diagram in α, β parameter plane. The remaining parameter was fixed at $\delta = 5$, and an initial point in phase space $X = 0.25$, $Y = 0.75$ was chosen. For each combination of α, β, the system was run for 5000 iterations after which the program checked for periodicities 1 through 15. The period number for a certain Arnol'd tongue can be retrieved in the column to the right in the proper colour code. Notably the tongues of low periods are very sparse in the figure, so it would be interesting to make a complementary picture of areas with positive Lyapunov exponents, which is quite easy

Bifurcations in the Utility Coefficient Plane

A different bifurcation diagram in parameter plane is produced in Fig. 8.10. Again we deal with the unit square, but now it is parameter space α, β and not phase space that is concerned. As we see the dominant shade is labelled 1, indicating fixed points. In the lower left corner there appears an irregularly concentric structure of periodicity "tongues" of a period adding appearance; 1, 2, 3, 4, 5, 6, with large gaps between indicating more complex dynamics.

8.5 Summary

To sum up, we suggested a unified model of asset price dynamics, the clue to which was the fact that assets or stock commodities, unlike flow commodities, remain on the market. Through trade these are redistributed among the agents, which, however, changes the basis for future plans and actions. Due to this unique demand/supply functions and market equilibria lose their meaning. If the system goes to equilibrium, there are infinitely many to choose from, and the one which it converges on depends on the dynamic process itself.

In the simple two agent two commodities model of Edgeworth type trade was limited to what the agent wanting to trade least from an actual asset distribution point was willing to exchange. In this way the agent wanting to trade more could get part of the way to higher utility, lacking possibilities to force the other to exchange more than he wants. Given such conditions for trade are assumed, it could happen that the dynamic trading process stuck at some disequilibrium asset distribution from which the system could not move further because both agents always wanted to sell or buy the same commodities. The excess demand driven tâtonnement process, overshooting unstable equilibrium points, could set up complex price dynamics, periodic or aperiodic.

A challenge to economists would be to set up a model with three or more agents. Then the intuitive Edgeworth model no longer serves, and it is not obvious which agents will remain most dissatisfied in a disequilibrium. This issue is not addressed in standard economics, though a solution is required even in macroeconomics for understanding how investment and saving always become equal through unintended inventory investments (unsold goods) and/or unintended saving (failure to buy commodities in shortage). Considering more than two agents would also make the tâtonnement process more natural and move the mind away from the case of bilateral monopoly.

Mathematically, though the model looks quite simple in statement; just a three-dimensional map with three branches of modest nonlinearity, it does not yield to any standard analytical methods. Even the simplest issue; the stability of equilibria is difficult to analyze. The equilibria are infinite in number, and the curve of equilibrium points in 3-dimensional state space is the intersection of three branch separating surfaces, so that the Jacobian cannot be uniquely defined at any equilibrium.[3] The model was first published in Puu (2015).

References

Edgeworth FY (1894) The pure theory of international values. Econ J 4:35–50

Puu T (1965) The effects of fiscal and monetary policy. Studia Oeconomia Upsaliensia 1:1–75

Puu T (2015) Disequilibrium trade and the dynamics of stock markets. In: Faggini M, Parziale A (eds) Complexity in economics: cutting edge research. Springer, Berlin, pp 225–245. ISBN 978-3-319-05185-7

[3]For this observation the author is indebted to his collaborator Dr Anastasiia Panchuk from the mathematics institute of NAS of Ukraine in Kiev.

Chapter 9
Macroeconomics and the Trade Cycle

Abstract In the following model we propose a simple change to the nonlinear Hicksian trade cycle model of 1950 through just internalizing capital stock. This brings no alien elements into the model, it just makes explicit what is there already, i.e., investment, considering that capital is the cumulative sum of successive investments. This makes it possible to tie the "floor" disinvestment to capital stock through its depreciation rate. The consequence is that one can dispense with the exogenous floor (constant, or growing) altogether. Through capital accumulation the model produces an endogenous growth trend, more explicitly, growth cycles around a trend. Thus also the Hicksian autonomous growth trend can be dispensed with, and the model becomes self contained. A problem then may seem to be that without these exogenous trends the growing variables, income and capital, cannot be reduced to stationarity through trend elimination. A new method, proposed by the author 55 years ago, which we call relative dynamics, replaces the growing income by the income growth factor and the growing capital by the capital to income ratio, and these appear as stationary time series, predominantly periodic The change removes arbitrary assumptions, such as equality of growth rates for the exogenous trends, in autonomous expenditures and the investment floor. This seems to be good as at second thought the floor level apparently must be decreasing rather than growing when capital accumulates. The change also produces both trend **and** cycles on its own, which the original multiplier-accelerator model cannot, and further reduces periodic growth rates from 50–100% in the original model to more realistic 0.2–10%.

9.1 Background

When the Great Depression spread around the industrialized world in the 1930s, existent economic theory had little understanding to offer. As a matter of fact this very theory was an obstacle to understanding. Its core was a system of interdependent markets in which the balance of supply and demand determined equilibrium prices. This was the idea of general equilibrium which went back to Léon Walras.

T. Puu, *Disequilibrium Economics*, https://doi.org/10.1007/978-3-319-74415-5_9

If, for instance, there was an excess supply of labour (involuntary unemployment), it would indicate that price, i.e., the wage rate, was too high to call forth sufficient demand. Different reasons could be given to such price rigidity, for instance any monopoly power of trade unions.

The idea that wages provided the substantial basis for household income, and that a cut, though making labour cheaper to hire, could also create a general deficiency of consumer demand, was there, but it was not part of economic theory. Such an effect was impossible to trace through the system of interdependent markets in a general equilibrium.

Several attempts were made to cope with this, and the most successful was by John Maynard Keynes (later Lord Keynes) in 1936 in his "*General Theory of Employment, Interest, and Money*".

The thrust was to think in entirely new categories, national income, consumption, investment, and saving. These concepts had been around in various contexts, but they had never been building blocks of theory before.[1] It also became essential to measure these as a basis for policy decisions. This was achieved through national accounting, modelled on common business accounting, so that any double counting of intermediate transactions was eliminated.

9.1.1 Flows

Basic in Keynesianism was the concept of periodic flow, of commodities or of money paid for these.[2] National income Y emerged through expenditures for investment goods I and for consumer goods C,

$$Y = C + I$$

and it was disposed of by expenditures on consumption C and saving S,

$$Y = C + S$$

As C is in both equations this implies that

$$I = S$$

[1] National income, for instance, according to Erik Lindahl "*Studies in the Theory of Money and Capital*" 1939, was considered as current interest on total wealth, itself the sum of the discounted values of all future incomes of every asset in society. A totally useless theoretical construct.

[2] Economic flow originally entered economics through Fracois Quenay's "Tableau Économique" in 1759 to illustrate goods and money circulating among social classes.

It had to be explained how this equality, sometimes called accounting identity, could always hold. The trick was to include inventories of unsold commodities in investment. In this way deficient demand could lead to accumulation of inventories, and be considered as involuntary investment. Likewise, shortage of commodities when not all consumers could get what they demanded could lead to involuntary saving.

But the equation could also be considered an equilibrium condition. This came in focus with the so called multiplier mechanism. A simple linear assumption concerning the relation between income, consumption and saving was that a fraction $C = cY$ was spent and the remaining fraction $S = (1 - c)Y$ saved. Using $I = S$ one could then solve for

$$Y = \frac{1}{1-c} I$$

Considering investment as given, any increment ΔI "multiplied up" income by the multiplier $\frac{1}{1-c} > 1$. The process could also be set up dynamically as a convergent series of secondary spendings. Seen in this context $S = I$ obviously was an equilibrium statement.

This dual interpretation lingered on, and it was never explained how involuntary investment and saving arise. The so called Stockholm School introduced the concepts "ex ante" and "ex post" for the components, but that hardly explained anything.[3]

The Keynesian macroeconomic model also contained a monetary element, as indicated in the title of Keynes's magnum opus. By tradition investment was assumed to depend on the rate of interest, as was saving, and it provided for the balance of investment and saving on the money market.

Keynes, however, dethroned saving to just a residual after consumption. As for investment, he argued that its sensitivity to interest was almost negligible, and if not, the rate of interest would anyhow be impossible to lower for stimulating investment in depressions, due to speculative forces on the money market.[4]

The true essence of macroeconomics became most clear in business cycle theory which will be our focus in this chapter, and it provided an entirely different explanation for investment through the principle of acceleration.

The idea of economic flows captured the minds for decades. Textbooks of the 1940s usually contained chapters about national accounting and pictures of the flows of commodities and money in the economy. We include a particularly attractive picture displaying flows in the Swedish economy tuned after actual national accounting data for 1947.

[3]Palander (1953) in his sole English publication, launches severe critique against the absence of explanations for transactions and pricing in disequilibrium, which might explain how entities such as involuntary investment are determined.

[4]Actually Keynes's reasoning about this is the most ingenious part of the work, but we have no need to enter the details, because the conclusion is that monetary factors are inessential.

9.1.2 The MONIAC

An amusing invention was the MONIAC (**MO**netary **N**ational **I**ncome **A**nalogue **C**omputer) hydraulic macroeconomic machine where flows of investment or export were represented by streaming coloured water in transparent tubes, and the stock of money by a water tank, whereas taxation was controlled through a valve.

It was invented by William Phillips in 1948, an Australian engineer studying at the London School of Economics. It was intended as pedagogical tool, but a screening session for prominent economists judged it as useful also for research. Then not only the London School of Economics, but every research environment that counted, such as Harvard Business School, must have one as well.[5]

Notably, the rather expensive physical machines had to be protected against overflow, which easily occurred with the linear relations involved. This was implemented through a number of floats and drains. One might imagine that this could have made economists think that such nonlinear protections against overflow might also have

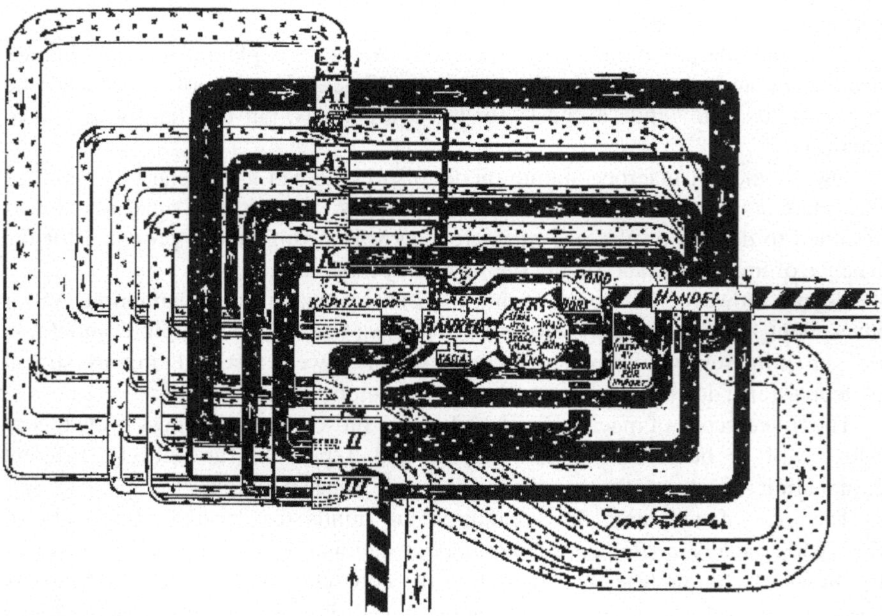

Fig. 9.1 Textbook illustration by Tord F. Palander from 1947 of the circulation of commodities and money in a Keynesian flow, adapted to actual Swedish national income data of that year

[5]The present author's supervisor, professor Tord Palander, who draw the picture Fig. 9.1, wanted a MONIAC also in Sweden, but he was outvoted in the research council by political scientists who claimed that if economists got an "economy machine" then they must also have a "state machine". As there was no such thing, the project collapsed.

some importance in the linear economies represented in the machines, but curiously it seems not to have been the case.

9.1.3 Protagonists

Keynes

Concerning John Maynard Keynes (1883–1946) we do not need to add much about his work, as it was discussed above. He spent his academic life in Cambridge, and was also born in the neighbourhood, son to economist John Neville Keynes. He was eventually nobilitated, not only knighted, but elevated to the House of Lords, and married to a famous Russian ballerina, Lydia Lopokova, though initially keeping his male lovers.

Samuelson

Paul A Samuelson (1915–2009), one of the most innovative and frugal economists of the twentieth Century, was an American economist of Jewish-Polish origin, eventually working at the MIT. Among his early contributions was "*Interactions between multiplier analysis and the principle of acceleration*" 1939 which provided the basis for the first self contained business cycle "machine", a linear second order oscillator. His interest for this area is natural as among his tutors were von Haberler, author of an almost exhaustive survey of business cycle theory 1937, and Alvin Hansen, author of Business Cycle Theory 1927.

Hicks

J.R. Hicks (1904–1989), a decade senior to Samuelson, was a British equivalent working at Oxford University. As mentioned his article "*Mr Keynes and the Classics: A Suggested Interpretation*" 1937, provided the most influential interpretation of Keynesian macroeconomics in terms of a precise model, and the slim monograph "*A Contribution to the Theory of the Trade Cycle*" 1950 extended Samuelson's linear business cycle model to a nonlinear format.[6]

[6]Hicks and Samuelson were competitors in a sense. In "*Foundations of Economic Analysis*" 1947 Samuelson improved the stability analysis in Hicks's "*Value and Capital*" 1939. In reverse, Hicks, as mentioned, in 1950 improved Samuelson's linear business cycle model to a non-linear format. The present author had the pleasure of meeting both, though Hicks wanted to discuss his interest in Portuguese Jesuit architecture in Western India, whereas Samuelson enquired about the author's supervisor Tord F. Palander, though not his scientific work but about his younger brother who was a well known medium who blindfolded could read unopened letters from unknown senders.

Among numerous other awards and distinctions, both were among the first to be awarded the so called Nobel Prize in economics, though in hindsight it seems that their reputations contributed more to the prestige of the prize than the prize contributed to theirs.[7]

9.1.4 The Multiplier-Accelerator Model

Quite as it is difficult to say where the idea of the multiplier originated, so is it difficult to trace the origin of the principle of acceleration.[8] The accelerator idea arose in the "overproduction" school of business cycles. See von Haberler (1937).

The idea is twofold: First, capital equipment is built up or let to decay in anticipation of changes in consumer demand. Second, as investments are related to expected changes of consumption, they tend to amplify or "accelerate" the process, hence bringing in a feature of instability. When multiplier and accelerator are linked together in one single feedback process, it is, of course, not possible to say whether investment has a lead over consumption or rather lags behind.

As formalized by Samuelson (1939), the principle says that investment is proportional to the rate of change of consumption. It is the natural companion to the multiplier, which relates consumption to investment, whereas the accelerator does the reverse. It also supplies the missing determinant for investment once the monetary factors in truly Keynesian spirit are scrapped.

Later Hicks (1950) realized that there is no need to restrict the action of the accelerator to changes in consumption expenditures alone, it should act in equal measure on all components of income, consumption, public spending, and investment itself. The difference in terms of model performance it marginal. As we will see below, Hicks also introduced other changes to the model, which, however, changed it radically into a nonlinear format.

---

[7]We write "so called" because it was not in Alfred Nobel's will. It appeared when the Bank of Sweden, one of the oldest central banks, had a 300 years anniversary. The bank had been a commercial bank as well and accumulated some handsome funds. Some people close to the government were also close to the academic discipline of economics, and they suggested a new prize. The Nobel Foundation fearing degradation refused. However, some time later they discovered they could earn much more on the stock market than on a bank account. However, the will stated that the funds be invested in the safe mode of a bank account. The government only could reinterpret Nobel's will, so the Foundation turned to the government. They answered, "yes, but on condition that you accept the prize", which was agreed. Later handsome sums of money have been offered to the Nobel Foundation for different new prizes, but they always refused.

[8]At least Aftalion in "*La réalité des surproductions génerales*" 1909, and J.M. Clark in "*Business acceleration and the law of demand*" 1917 seem to have had a clear idea of it.

Impact of Hicks's Model

The multiplier-accelerator model in Hicks's nonlinear format made an enormous impact around 1950. It was in a sense the mere essence of mathematical economics. At a conference about it Sir Dennis Robertson (1890–1963), close collaborator to Keynes in Cambridge and leading explorer of the multiplier, contributed a poem which the present author cannot withhold the reader.

A Poem

As soon as I could safely toddle
My parents handed me a model.
My brisk and energetic pater
Provided the accelerator,
My mother, with her kindly gumption,
The function guiding my consumption;
And every week I had from her
A lovely new parameter,
With lots of little leads and lags
In pretty parabolic bags.
With optimistic expectations
I started on my explorations,
And swore to move without a swerve
Along my sinusoidal curve.
Alas! I knew how this would end;
I've mixed the cycle and the trend,
And fear that, growing daily skinnier,
I have at length become non-linear.
I wander glumly round the house
As though I were exogenous,
And hardly capable of feeling
The difference 'tween floor and ceiling.
I scarcely now, a pallid ghost,
Can tell ex ante from ex post;
My thoughts are sadly inelastic,
My acts incurably stochastic.

By Sir Dennis Robertson 1950. He was so delighted with this himself that he published it twice.[9]

[9]This poem, first read at a British national conference about the Hicksian model illustrates how important the model was considered, in a sense the very quintessence of mathematical economics. Notable is also how clear the audience was about the nature of nonlinearity created by floor and ceiling. To make the poem justice it should be read loud with appropriate mispronunciations, such as "exogenaus".

9.2 The Models

The scope of the present chapter is to make a few amendments to the Samuelson–Hicks model, starting from the Hicksian nonlinear version with floor and ceiling. As we remember, the original Samuelson model is linear, so Hicks's innovation provides two bounds, that force the system to move within the space between them. We do this in two steps, first "floor", as we like Duesenberry (1950) feel it is both the most important element and sufficient by itself for getting persistent bounded oscillations.

We cite R.G.D. Allen's (1956) very clear account of it: "*On pursuing this point as Duesenberry does, it is seen that the explosive nature of the oscillations is largely irrelevant, and no ceiling is needed. A first intrinsic oscillation occurs, the accelerator goes out in the downswing, and a second oscillation starts up when the accelerator comes back with new initial conditions. The explosive element never has time to be effective - and the oscillations do not necessarily hit a ceiling.*"

The floor is activated when the principle of acceleration would make disinvestment exceed the natural depreciation that occurs when no worn out capital at all is replaced.

The bounds in the Hicksian version are imposed exogenously, as is a growth trend for "autonomous" expenditures (such not explained through the internal forces of the multiplier or accelerator). This model itself can explain *either* growth or oscillations, not oscillations *and* growth. To reduce the resulting model to something that can be analyzed by standard methods, all three trends are assumed to grow at the same rate, this is what Hicks suggested, and also what everybody (Rau, Gandolfo, Hommes, Gallegati) analyzing the model did.

The assumption of equal rates is so arbitrary that one may wonder if it is admissible even in theory. Further, to the present author's intuition there is a serious slip in Hicks's argument. According to his own reasoning the floor is related to maximum depreciation on capital. But when the accelerator is in action, which must be during most of the cycle, then investments are positive, and capital stock is accumulating. This means that the possible maximum depreciation *increases* with time as more capital is invested, which obviously would allow for disinvestments *larger* in absolute value. Hence the floor would rather be decreasing, i.e., go the other way than the autonomous growth trend and the ceiling.

This in particular makes it illicit to assume equal *growth* rates for all trends.

However, we avoid all these problems through just making capital endogenous in the model. To do this, we do not even need to add any assumptions that are not there already. Capital is just the cumulative sum of investments, and there is a theory for investment in the model. It is either positive when the principle of acceleration works in booms, or negative when the principle is suppressed in slumps. So, why not define this cumulative sum of investments as capital stock?

The first advantage then is that we can assume a rate of capital depreciation, and formally relate the floor to the stock of capital. In this way we can dispense with the floor as an exogenous bound.

We can also skip all the other exogenous trends. So, how about economic growth without autonomous expenditures? As a matter of fact we do no longer need this trend as the modified model produces a growth trend by itself, more precisely, growth rate cycles around a trend. This is a second advantage of including capital (as it raises the degree of the system from second to third).

There seemingly remains one problem; how can we get a stationary time series to analyze when there are no growth trends to eliminate? In this connection the present author recalled that he 55 years ago published a little note in Oxford Economic Papers, Puu (1963), about the original multiplier/accelerator model, as a pedagogical device for those economics students who had not learned complex numbers to understand the closed form solution of a second order oscillator.

In reduced form the model read $Y_t = (a + c)\, Y_{t-1} - aY_{t-2}$. Through defining $y_t = \frac{Y_t}{Y_{t-1}}$, one simply got $y_t = (a + c) - \frac{a}{y_{t-1}}$, a nonlinear but first order iteration that could be displayed on a graph like those now always used to illustrate nonlinear iterations.[10]

In this context the idea, however trivial, comes to good use; also the version with capital can be converted in the same way. The present author and collaborators proposed to call this conversion "relative dynamics". This produces the stationary series that one needs for analysis, so we can do without the trends also for analysis.[11]

9.2.1 *The Original Samuelson–Hicks Model*

Let us briefly recapitulate the original Samuelson–Hicks model. It was based on two interacting principles: consumers, consuming a fraction c of past income, $C_t = cY_{t-1}$, and investors aiming at maintaining a stock of capital K_t in given proportion a to income Y_t. With an additional time lag for the construction of capital, net investments, by definition the change in capital stock, $I_t = K_t - K_{t-1}$, become $I_t = a\,(Y_{t-1} - Y_{t-2})$.

As income is generated by consumption and investments, $Y_t = C_t + I_t$, a simple feedback mechanism

$$Y_t = (c + a)Y_{t-1} - aY_{t-2}$$

was obtained. As mentioned it could generate growth or oscillations in income.

[10]But at that time it was even difficult to publish, it was rejected by at least Journal of Economic Theory and Review of Economic Studies, as being trivial until the Oxford journal showed mercy.

[11]The model was proposed repeatedly in the period 2004–2007, but nobody seemed to be interested. Maybe the present author is to blame as the first paper, written with two mathematicians, went under the name: "Tongues of periodicity in a family of two-dimensional maps of real Möbius type".

To be quite true to history, Samuelson applied the accelerator to consumption expenditures only, whereas Hicks applied it to all expenditures. The difference in terms of model behaviour is negligible.

However, there was more to the Hicksian reformulation. He realized that there must be limits to the accelerator-based investment function. In a depression phase $I_t = a(Y_{t-1} - Y_{t-2}) < 0$, so it can even happen that income (=production) decreases at a pace so fast that more capital can be dispensed with than what disappears through natural wear. As nobody actively destroys capital, there is a lower limit to disinvestment, the so called "floor", fixed at the (negative) net investment that occurs when no worn out capital at all is replaced.

At the same time Hicks suggested that there be a "ceiling" at full employment, when income could not be expanded any more. Hicks never assembled the pieces to a complete formal model. It is clear that the floor constraint is applied to the investment function, so it becomes something like

$$I_t = \max\left(a\left(Y_{t-1} - Y_{t-2}\right), -I_t^{\text{floor}}\right) \tag{9.1}$$

where I_t^{floor} is the absolute value of the floor disinvestment.

On the other hand it is not quite clear what the ceiling is applied to. Most likely Hicks thinks of it as applied to income, so that the income formation equation is changed to

$$Y_t = \min\left(cY_{t-1} + I_t, Y_t^{\text{ceiling}}\right) \tag{9.2}$$

where Y_t^{ceiling} is the full employment capacity income.

Gandolfo (1985) interpreted the model this way, and Hommes (1991) produced a more or less full analysis. It seems that the first formalization in this format is due to Rau (1974).[12]

To the complete model also belong the autonomous expenditures. These can be constant, or growing. In his verbal description Hicks seems to have been in favour of exponentially growing autonomous expenditures, pictured as straight lines in a semi-logarithmic graph. Obviously he wanted growth and cycles around a growing trend. Growth, however, was not created endogenously when it produced cycles, the choice depended on the parameters, i.e., whether the eigenvalues were real or complex. If both were wanted the growth trend must be introduced exogenously.

[12]It is not obvious which agents cut their expenditures when the ceiling is reached. As an alternative one might incorporate it in the investment function, along with the floor, thereby implying that it is the investors who cut further investment when they realize that full employment is reached. This was the choice of Goodwin (1951) and many other students of the Hicksian business cycle machine, including the present author (see Puu 1987; Sushko et al. 2004).

To make this type of model suitable for analysis, the floor and ceiling must then be growing too, and at the same rate as the autonomous expenditures. This seems to have been Hicks's own tacit assumption. The assumption of equal growth rates is, of course, arbitrary, and, as we saw not logical as far as the floor is concerned.

Gandolfo modelled it this way, though Hommes preferred to analyze the stationary case where autonomous expenditures, floor, and ceiling were all constant.[13]

9.2.2 Digression on Technology

It is appropriate to add a little digression here about the kind of production technology that is assumed as lying behind the principle of acceleration.

The most obvious candidate would be a production function of the Leontief fixed proportion type,

$$Y = \min\left(\frac{K}{a}, \frac{L}{b}\right)$$

where K and L are capital and labour, Y is real income produced, and a, b are given coefficients representing the production technology. Given this the firms would build up a capital stock equal to $K = aY$ and hire a labour force equal to $L = bY$, neither more nor less. Observe that we did not date the variables yet. We will do this in two different ways below.

(i) As long as we are concerned with the accelerator, Y should represent expected production $\widetilde{Y}$, and $\widetilde{K}$ and $\widetilde{L}$ the planned capital stock and labour force respectively. We further have to recognize that Y is a flow variable (attributed to time *periods*), whereas K and L are stock variables (attributed to time *points*), for instance the starting or ending points of periods. Accordingly a convention for how points and periods fit together is needed. This is, of course, arbitrary, so we can decide that time period t begins with time point $t-1$ and ends with time point t.

Given we assume that for a certain period any planned production would need the appropriate capital and labour to be available from the beginning of that period we would write:

$$\widetilde{Y}_t = \min\left(\frac{\widetilde{K}_{t-1}}{a}, \frac{\widetilde{L}_{t-1}}{b}\right)$$

Assume now that we have a naïve forecasting rule: $\widetilde{Y}_t = Y_{t-1}$, and that the formation of capital takes one period, i.e. $\widetilde{K}_{t-1} = K_t$, as does hiring labour

[13] A later mathematical analysis of Gandolfo's case may be found in Gallegati et al. (2003).

force, i.e. $\widetilde{L}_{t-1} = L_t$. Then we have:

$$Y_{t-1} = \min\left(\frac{K_t}{a}, \frac{L_t}{b}\right)$$

whence $K_t = aY_{t-1}$ and similarly $K_{t-1} = aY_{t-2}$, provided capital is the limitational. input.

Due to definition, net investments become

$$I_t = (K_t - K_{t-1}) = a\,(Y_{t-1} - Y_{t-2})$$

which justifies the principle of acceleration.

(ii) Anticipating the Hicksian ceiling, we can also interpret the production function differently, as a capacity limit determining actual maximum production in a period as dependent on actual capital and labour available in its beginning, i.e.,

$$Y_t^{\text{ceiling}} = \min\left(\frac{K_{t-1}}{a}, \frac{L_{t-1}}{b}\right)$$

When capital is limiting, then $Y_t \le Y_t^{\text{ceiling}} = \frac{K_{t-1}}{a}$ must hold. In case the process actually sticks to the ceiling, even $Y_t = \frac{K_{t-1}}{a}$. This may seem contradictory to and reversing causality in the statement under (i), but it is not! The ceiling and the accelerator never operate at the same time, and they are based on different logic. In Hicks's original argument it seems that he is in favour of labour shortage providing the ceiling, though to keep the model to the two endogenous variables, income and capital, we will explore whether capital itself may be used also for the ceiling limit.

This digression may seem a bit pedantic, but it aims at clearing up the issue for the meticulous reader.

9.3 Floor Only

9.3.1 The Formal Model

Recall the original reduced form equation

$$Y_t = a\,(Y_{t-1} - Y_{t-2}) + cY_{t-1} \tag{9.3}$$

The conversion to relative dynamics, worked out in Puu (1963), simply defined $y_t = \frac{Y_t}{Y_{t-1}}$, which is the ratio of incomes in two subsequent periods, so that $y_t - 1$ denotes the rate of growth. Dividing (9.3) through by Y_{t-1} and using the new variable, we obtain

$$y_t = (a + c) - a\frac{1}{y_{t-1}} \tag{9.4}$$

which, as we see, is first order, though nonlinear.

The Absolute Growth Dynamics

Let us first just restate the consumption function

$$C_t = cY_{t-1} \tag{9.5}$$

Let r denote the rate of capital depreciation, and put $I_t^{\text{floor}} = rK_{t-1}$. The investment function then becomes

$$I_t = \max\left(a\left(Y_{t-1} - Y_{t-2}\right), -rK_{t-1}\right) \tag{9.6}$$

c, a and r are parameters such that $0 < c < 1, a > 0, 0 < r < 1$. The last condition for the depreciation rate may seem less obvious than that the propensity to consume is in the unit interval, but, as we are in discrete time dynamics anything else would mean that total capital stock could be annihilated in one time period (which would be quite like a rate of interest exceeding -100% which converts deposits to debts and vice versa).

We now also need a relation for capital stock updating

$$K_t = K_{t-1} + I_t \tag{9.7}$$

which just says that capital stock changes with net investments according to (9.6), accelerator generated as

$$K_t = K_{t-1} + a\left(Y_{t-1} - Y_{t-2}\right)$$

or, in the case when the floor is activated, just decays, like a radioactive substance

$$K_t = (1 - r)\, K_{t-1}$$

As there is no ceiling, the income generation equation reads

$$Y_t = C_t + I_t \tag{9.8}$$

Equations (9.5)–(9.8) now define the complete system. It is easy to see through numerical studies that the model can create a process of accumulating capital, along with a growth trend in income, and this without any growing autonomous expenditures at all. Further, the model creates growth cycles around these secular trends. This model first appeared in Puu et al. (2005).

Fixed Points and Their Stability

Let us first note that if

$$a(Y_{t-1} - Y_{t-2}) + rK_{t-1} \geq 0 \tag{9.9}$$

then the first alternative in (9.6) applies. Let us call the part of phase space where (9.9) is satisfied Region I. Then, eliminating the consumption and investment variables through substitution from (9.5) into (9.7) and (9.8), we see that in Region I the system is defined by

$$K_t = K_{t-1} + a(Y_{t-1} - Y_{t-2}) \tag{9.10}$$

$$Y_t = cY_{t-1} + a(Y_{t-1} - Y_{t-2}) \tag{9.11}$$

Let us now look at the other alternative, where the second branch of (9.6) is activated (i.e., $I_t = -rK_{t-1}$). This occurs when

$$a(Y_{t-1} - Y_{t-2}) + rK_{t-1} < 0 \tag{9.12}$$

Let us call this Region II. From (9.5) to (9.8) we then get

$$K_t = (1 - r)K_{t-1} \tag{9.13}$$

$$Y_t = cY_{t-1} - rK_{t-1} \tag{9.14}$$

It is easy to find the fixed points for (9.10)–(9.11). From (9.11), there is just one fixed point for income, $Y_t = Y_{t-1} = Y_{t-2} = 0$. Next, putting $Y_{t-1} = Y_{t-2}$ in (9.10), we conclude that $K_t = K_{t-1} = K$, i.e., any (positive) capital stock may be an equilibrium stock. Numerical experiments indicate that, depending on the dynamical process (i.e., on the initial conditions), the stock of capital may end up at different equilibrium values.

As for stability, this also implies that, if there is some perturbation of the capital stock, then the process will eventually end up at some new equilibrium stock. However, income always goes to the single zero equilibrium. This, of course, is true only if the equilibrium is stable.

It is also obvious that the system (9.13)–(9.14) has only one fixed point located at the origin of phase space.

Let us first investigate the stability of the system (9.10)–(9.11). As (9.11) is independent of the capital stock, we can study this single equation alone. However, we have to observe that (9.11) is a linear second order difference equation. Writing down the Jacobian matrix of (9.11) and the corresponding characteristic equation, one can easily get its eigenvalues:

$$\lambda_{1,2} = \frac{1}{2}(a + c) \pm \frac{1}{2} \tag{9.15}$$

From (9.15) we can immediately see that the zero fixed point is a node if $(a + c)^2 > 4a$ and a focus if $(a+c)^2 < 4a$. It is stable iff $\left|\lambda_{1,2}\right| < 1$. The latter condition holds for the parameter ranges: $c < 1$, and $-\frac{1+c}{2} < a < 1$. Taking account of the feasible parameter range, we conclude that the stability region for the fixed point of (9.11) is $0 < c < 1, 0 < a < 1$.

Because of the linearity, the system (9.10)–(9.11) is a contraction for that parameter range, not only at the fixed point, but in the entire Region I.

The system (9.10)–(9.14) becomes an expansion if $|\lambda_{1,2}| > 1$, which happens for $a > 1$.

Now consider the stability of (9.13)–(9.14). Its eigenvalues are $\mu_1 = c$ and $\mu_2 = 1 - r$. As $0 < r < 1$ and $0 < c < 1$, we note that both eigenvalues are positive and less than unity. Accordingly, the system (9.13)–(9.14), defined in Region II, is a contraction.

The economics of this is that in Region II the system tends to equilibrium, with zero capital and zero income. The stability of the map (9.13)–(9.14) would be a problem for the model if the process were not mapped back into Region I, in which the fixed point may be unstable. The process can jump back and forth between the Regions any number of times, so being kept going forever. If $r < 1 - c$, $(a + c)^2 < 4a$, and $a > 1$ hold, then this jumping between regions occurs in finite numbers of iterations for any initial conditions we may care to choose.

The first condition states in terms of subject matter that the rate of capital depreciation r be less than the rate of saving $1 - c$. This seems to be a condition fairly likely to be fulfilled in most cases.

The Fixed Point Bifurcation

As we have seen, only the fixed point of (9.11) may become unstable. Indeed, at

$$a = 1 \tag{9.16}$$

the eigenvalues (9.15) are complex conjugates and have unit modulus. Thus, the fixed point has a bifurcation analogous to the Neimark bifurcation. At the bifurcation we can write the eigenvalues (9.15) as

$$\lambda_{1,2} = \cos\theta \pm i \sin\theta$$

where

$$\cos\theta = \frac{1}{2}(a + c) \tag{9.17}$$

and

$$\sin\theta = \frac{1}{2}\sqrt{4a - (a + c)^2} \tag{9.18}$$

In the case there is a rational rotation around the fixed point with rotation number $\frac{m}{n}$, the solution is $\theta = 2\pi\frac{m}{n}$. Then, using (9.17) and (9.16), we get the exact value of parameter c which corresponds to the rotation number $\frac{m}{n}$, i.e.,

$$c = 2\cos\left(\frac{2\pi m}{n}\right) - 1 \tag{9.19}$$

For instance, at the bifurcation (9.16) the rotation number $\frac{m}{n}$, where $m = 1$ and $n = 1, \ldots, 6$ holds for $c = 1, -3, -2, -1, \frac{\sqrt{5}-3}{2}$ and 0, respectively. Obviously, these values are not admissible, so $\frac{m}{n} = \frac{1}{6}$ is the lowest basic resonance that falls into the admissible parameter range $0 < c < 1$, starting at $c = 0$, $a = 1$. We see the points c according to (9.19) for $n = 6, \ldots, 15$ marked by the ascending sequence of circles on the $a = 1$ line in Fig. 9.4.

Our study of this bifurcation was confined to the fixed point of (9.10)–(9.11), and the reader may note that then (9.11) is no different from the original Samuelson–Hicks model.

Eventually it is the growing capital stock that is responsible for the secular growth created by the proposed model for $a > 1$.

In Fig. 9.2, we show typical growth paths for capital and income. Income is the curve with the reddish support, oscillating around the equilibrium value, with increasing height of the peaks and increasing amplitude. Notably we could only

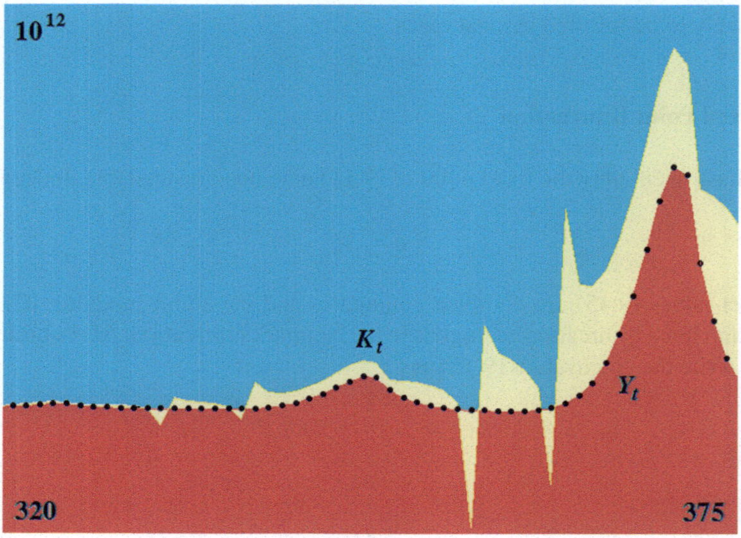

Fig. 9.2 Traces of capital and income for the model over a small number of iterations when parameters are $a = 2.25$, $c = 0.65$, $r = 0.01$. Obviously income is growing with short recessions, and so is capital with dips while it is left to depreciate at the floor level. Note that both the scales and the averages for capital and income are different in order to accommodate both traces in the same picture. The vertical scale refers to the income curve. The periods are marked by dots on the income trace. As we will see after variable transformation there is a 23 period growth cycle, though this is impossible to infer from the growing trace

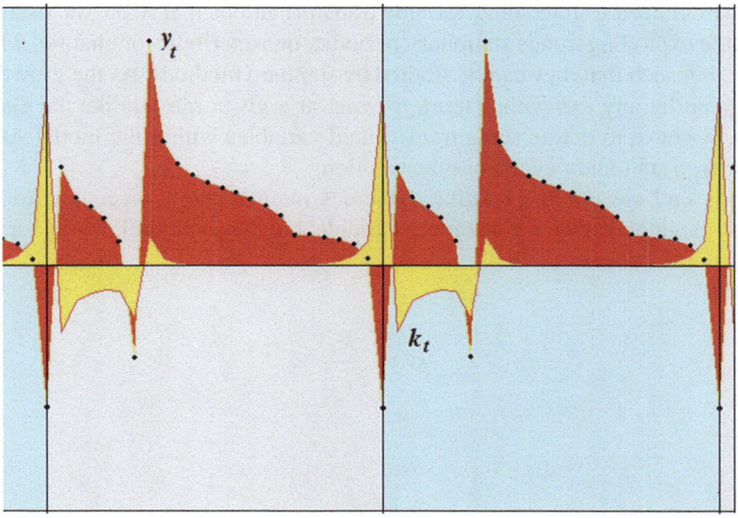

Fig. 9.3 The relative dynamics traces transformed from the preceding picture. These clearly show up a 23-period cycle. The parameters are as before $a = 2.25$, $c = 0.65$, $r = 0.01$. The cycle results in a growth factor 7.4 over the entire cycle, which, averaging geometrically over the cycle means 1.09 (9%) per period. For comparison, when the roots are real for the original multiplier-accelerator model we can calculate a growth factor of $\sqrt{2.25} = 1.5$ (50%) per period, which would result in $1.5^{23} = 11{,}223$ over the entire cycle

display 55 iterations, as otherwise any curve would be horizontal at the left end and leave any frame at the right.

The picture was calculated for parameters $a = 2.25$, $c = 0.65$, $r = 0.01$. As we see in Fig. 9.3, this case results in a 23-period cycle around a growth trend. As the growth *rate* makes a constant amplitude oscillation, income itself oscillates with increasing amplitude.

As for capital, the curve with yellow support shows an ever increasing trend with periodic dips. As we see, the slope of these increases. This is a reflection of the fact that the floor restriction slackens with increasing capital, and larger disinvestment is allowed as capital stock grows. The falling segments occur where the floor is activated.

In Fig. 9.3 for the relative dynamics of the case, we indicate the different phases; as for Fig. 9.2 we cannot do this, because it is impossible to detect any periodicity in the growing traces.

9.3.2 Stationary Relative Dynamics

It is convenient to study the phenomena in absolute dynamics only when $a \leq 1$, that is when the fixed point for income is stable. If not, then there is exponential growth in the model, so all variables eventually explode.

We would need to find some variable transformations that make the oscillations around the exploding trends stationary periodic, quasiperiodic, or chaotic, whatever they are, but such that they can be studied by standard methods. As the growth trend is not given by any exogenous term growing at a given rate, unlike the Gandolfo version, we have to define some transformed variables within the model such that they undergo stationary cyclic or other motion.

To this end we use the relative dynamics method, the conversion from (9.3) to (9.4) though we now have one more variable, but the same trick works for capital as well.

Define

$$k_t = \frac{K_t}{Y_{t-1}} \tag{9.20}$$

and

$$y_t = \frac{Y_t}{Y_{t-1}} \tag{9.21}$$

These new variables are the *actual* capital/output ratio in any phase of the cycle, as distinguished from the optimal ratio a, and the relative change of income from one period to the next. Again there is a reduction of dimension for the system, now from 3 to 2. Using these new variables defined in (9.20)–(9.21),we can restate the dynamical system as follows.

Suppose we have

$$k_{t-1}(a(y_{t-1} - 1) + rk_{t-1}) \geq 0$$

This corresponds to Region I in the original model as we see from (9.9). The reader may wonder about the occurrence of k_{t-1} as a multiplicative factor in this new region definition.

The reason for this is that division of the inequalities (9.9) or (9.12) through by Y_{t-1}, which can take on a negative sign, may change the sense of the inequality. For this reason we would have to split the two regions in four, but we can avoid this complication if we multiply the left hand sides of the inequalities (9.9) or (9.12) through by a variable defined in the new system (9.20)–(9.21), which always takes the sign of Y_{t-1}. Such a variable is k_t, as we see from (9.20), because capital stock K_t is always nonnegative.

Using (9.20)–(9.21), the system in Region I can then be written

$$k_t = \frac{k_{t-1}}{y_{t-1}} + a\left(1 - \frac{1}{y_{t-1}}\right) \tag{9.22}$$

$$y_t = c + a\left(1 - \frac{1}{y_{t-1}}\right) \tag{9.23}$$

Suppose that, on the contrary,

$$k_{t-1}(a(y_{t-1}-1)+rk_{t-1}) \geq 0$$

This obviously corresponds to Region II in the original model. Then (9.22)–(9.23) are replaced by

$$k_t = (1-r)\frac{k_{t-1}}{y_{t-1}} \tag{9.24}$$

$$y_t = c - r\frac{k_{t-1}}{y_{t-1}} \tag{9.25}$$

Fixed Growth Points

Written as relative dynamics (9.24)–(9.25), the new system as well has fixed points. In terms of economics they represent equilibrium growth rates. Consider first Region I. Putting $k_t = k_{t-1} = k$, $y_t = y_{t-1} = y$ in (9.22)–(9.25), we obtain

$$k = a \tag{9.26}$$

and

$$y^2 - (a+c)y + a = 0 \tag{9.27}$$

According to (9.26) the equilibrium capital/output ratio k equals the optimal one as indicated by the accelerator a, which intuitively seems most reasonable. As for (9.27), it determines either two real or two complex conjugate equilibrium values for the relative income growth rate

$$y_{1,2} = \frac{1}{2}(a+c) \pm \frac{1}{2}\sqrt{(a+c)^2 - 4a} \tag{9.28}$$

We may observe that (9.38) is equal in form to the characteristic equation for the original Samuelson–Hicks model, so the growth rate equilibria exist whenever the original multiplier-accelerator model has solutions with two real roots.

A saddle-node bifurcation resulting in (dis)appearance of the real fixed points (k, y_1) and (k, y_2) obviously occurs when $(a+c)^2 = 4a$.

For $(a+c)^2 < 4a$ no fixed point exists for the income growth rate, whereas for $(a+c)^2 > 4a$ there are two fixed points, one stable and one saddle. This is easy to see from (9.23), as the derivative (putting $y_{t-1} = y$ at equilibrium) is $\frac{\partial y_t}{\partial y_{t-1}} = ay^2$.

Using the larger root according to (9.28), we get

$$\frac{a}{y_1^2} = \frac{(a+c) - \sqrt{(a+c)^2 - 4a}}{(a+c) + \sqrt{(a+c)^2 - 4a}} < 1 \tag{9.29}$$

and, using the smaller

$$\frac{a}{y_1^2} = \frac{(a+c) + \sqrt{(a+c)^2 - 4a}}{(a+c) - \sqrt{(a+c)^2 - 4a}} > 1 \tag{9.30}$$

So far we only checked the stability of (9.28). We should complete the discussion by differentiating (9.22)–(9.23) and again deleting the index in the right-hand side,

$$\frac{\partial k_t}{\partial k_{t-1}} = \frac{1}{y} \tag{9.31}$$

so stability for the capital variable at the fixed point depends on the equilibrium value of y. From (9.28) we see that if the roots are real, then both are positive (due to the minus term under the root sign). Further, whenever $c < 1$, which it must be to make any sense, we have $1 < y_2 < y_1$ according to (9.28). Hence,

$$\frac{1}{y_1} < 1 \tag{9.32}$$

and

$$\frac{1}{y_2} < 1 \tag{9.33}$$

The fixed point (k, y_1) is a stable node whereas (k, y_2) is a saddle. This is a somewhat pedestrian way of checking stability, for income and capital separately, but, quite as in the original model, the Jacobian matrix of the system (9.22)–(9.23) is triangular, making the main diagonal derivatives eigenvalues.

Fixed Decline Points

As was the case before the reduction to relative dynamics, the model also has fixed points in Region II (i.e., where (9.24)–(9.25) hold). Put again $k_t = k_{t-1} = k$, $y_t = y_{t-1} = y$ in (9.24)–(9.25). The system obviously has a fixed point where

$$k = \frac{1-r}{r}(c + r - 1) \tag{9.34}$$

and

$$y = 1 - r \tag{9.35}$$

If we recall that the new y-variable is the income ratio for one period to the previous, then we realize that the fixed point means relative income decrease at the constant rate $(1 - r)$, which is the rate of capital depreciation. The accelerator then results in the same investment as the floor condition, and capital is just depreciating at the same rate. When the system is in this new fixed point, corresponding to a negative constant growth rate $-r$, then the system is on its way towards the zero equilibrium for capital and income according to the original setup in Region II. The eigenvalues in the fixed point are 0 and $\frac{c}{1-r}$. It is stable if $r < 1 - c$, but for the same condition this fixed point does not belong to Region II. Thus, the system switches between the two definition regions, so the attractivity of the new fixed point does not really matter.

There is an additional fixed point in Region II:

$$k = 0 \tag{9.36}$$

and

$$y = c \tag{9.37}$$

The eigenvalues are 0 and $\frac{1-r}{c}$, so as the second eigenvalue is the reciprocal to that in the previous case, it exchanges stability with the aforementioned

9.3.3 *Periodicity*

So far we studied the fixed points of the relative dynamics system. Numerical experiment indicates cycles in the growth rates. For instance, as we see in Fig. 9.3, the parameter combination $a = 2.25$, $c = 0.65$, $r = 0.01$ results in a 23-period growth cycle. The case was illustrated in Fig. 9.2, though it was impossible to see any regular pattern of oscillation in that picture. In Fig. 9.3 the periodicity is easy to identify, as we marked the successive iterates with little dots, it is easy to count them. In the figure we can identify slightly more than two full cycles separated in grayish and bluish zones.

The trace with red support is for the capital to income ratio. It is worth noting the long horizontal sections in which the ratio equals the accelerator coefficient, which seems to catch the trajectory for considerable periods of time.

To see some more cases, we show a bifurcation diagram in Fig. 9.4. It represents the parameter plane a (horizontal), c (vertical). The third parameter r is fixed at 0.01.

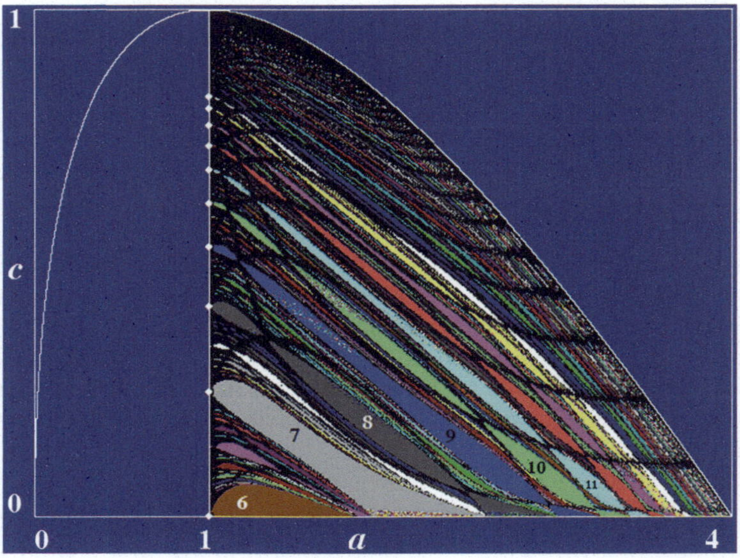

Fig. 9.4 Bifurcation diagram showing tongues of periodicity. The periods are calculated for the relative dynamics model, income growth ratio and capital income ratio. The income and capital variables themselves are growing, and cannot be analyzed by methods for stationary time series. Notable is that periodicity is now main frame, whereas in the original multiplier-accelerator model it was quasiperiodicity. Again we have the twisted tongues due to shifts in regime application (accelerator versus floor), producing the curious pattern of black streaks. The latter seem to call for closer examination

Changing this parameter, the rate of depreciation, changes little in the picture. Such changes mainly affect the structure of black streaks that seem to run through the points where the periodicity tongues look twisted.[14]

The tongues were computed for the relative dynamics model (9.22)–(9.25), checking periodicities 1–45, quite as described in Chap. 3. However, the bifurcation points on the line $a = 1$ where the periodic tongues start were computed using (9.18), as explained above. On the bifurcation line, the original system is periodic (for rational rotation numbers), so this is not surprising. To the left of the vertical line at $a = 1$, the zero fixed point of the original system is stable.

In Fig. 9.4 we also see a parabola turned upside down. It is the locus of points $(a+c)^2 = 4a$, the borderline between complex (below) and real (above) roots to the characteristic equation. To the left of the line $a = 1$, the attracting zero fixed point is a focus below the parabola and a node above it. To the right of the line $a = 1$, we have the tongues of periodicity drawn below the parabola. Above it there is the area where the roots (9.28) are real, and there is an attractive fixed point for the relative

[14]For the mathematician this effect might be interesting to study as the twists in the periodicity tongues seem to occur where the order of applications of region regimes is changed.

dynamics system. The original system then goes to a stable growth rate at the larger value of (9.28).

9.3.4 Absolute and Relative Variables

We should now clarify a little more the relation between the 3D map in original absolute variables $(K_{t-1}, Y_{t-1}, Y_{t-2}) \longmapsto (K_t, Y_t, Y_{t-1})$ and the derived 2D map in relative variables $(k_{t-1}, y_{t-1}) \longmapsto (k_t, y_t)$. The need for the relative system arises in the case when the accelerator generated motion is explosive (i.e. when $a > 1$ holds), which, of course, is the interesting case. If so, then the absolute variables always explode to infinity.

This can occur in two ways: when growth rates $y_{1,2}$ according to (9.28) are real, then asymptotically the absolute system settles to exponential growth at the higher (stable) rate. If they are conjugate complex, which is the more interesting case, then the relative system settles at oscillatory motion in the growth rates. As we saw in the bifurcation diagram Fig. 9.4, this oscillatory motion is predominantly periodic.

The location of the parameter point that our example displayed in Figs. 9.2 and 9.3, is within the 23-period tongue and is difficult to detect because the tongue is so tiny and stuffed among other tongues.

Easier to find are parameter points such as $a = 1.25$, $c = 0.05$, in the six-period tongue, or $a = 1.25$, $c = 0.25$, located in the seven-period tongue, both leading to regular motion of low periodicity. Unfortunately these low periodicities correspond to parameter points for which the propensity to consume is unrealistically low, but they may yet be useful for further numerical studies.

The periodicities are a novel feature of the present model when the floor is tied to the stock of capital. In the original Samuelson–Hicks model, $Y_t = (c + a)Y_{t-1} - aY_{t-2}$, periodic solutions do *not* occur. The closed form solution is a product of a growth factor ρ^t, where $\rho = \sqrt{a}$, and an oscillatory factor, $\cos(\theta t)$, where $\theta = \arccos\left(\frac{a+c}{2\sqrt{a}}\right)$.

It is extremely unlikely that $\theta = 2\pi \frac{m}{n}$ (i.e., that the oscillation frequency is a rational multiple of 2π), so oscillatory motion is always quasi periodic. In the (a, c)-parameter plane, $\cos\left(\frac{2\pi m}{n}\right) = \frac{a+c}{2\sqrt{a}}$ for m, n integers then produces a set of thin curves, not a set of thick tongues with nonzero area measure.

As we saw, the periodic growth factor for this original model is $\sqrt{a}$. Numerical studies indicate that in our model, the growth factors are considerably smaller, even if there still is exponential growth as long as $a > 1$ holds (see below for a numerical example).

For the case $a < 1$, with a stable zero fixed point for income, we still need the model in absolute variables because the relative system then makes no sense.

However, even with explosive motion we are still primarily interested in the absolute variables, income rather than its growth rate (note that motion of the relative variables is always bounded, even when the absolute variables explode).

It is then worth noting that the original system for the evolution of capital and income can always be retrieved from any solution to the relative system. One could arbitrarily put income in the first time period equal to unity and then obtain its evolution as a continued product of the growth factors calculated from the relative model. A different initial first period income would then just scale the entire time series obtained (up or down) in proportion.

What has been said also means that we can regenerate all further cycles in the time series from the values for just one cycle through multiplication by this continued product over one cycle. Note how this lowering of growth rates in the present model, as compared to the original Samuelson–Hicks model, strengthens the Duesenberry argument, as it is now much less likely that the present moderate growth rates will ever hit any "ceiling" growing with labour force or the like.

If the solution for the growth rates is periodic, then the continued product over a complete cycle takes a given constant value, no matter at which observation we start. This constant for our exemplifying 23-period cycle is for instance 7.400966..., which, averaging geometrically over a complete cycle, yields a growth factor of 1.0909... per period for the income variable, i.e., 9%, still high but close to realistic. This can be compared to the growth factor for the original Samuelson–Hicks model, which equals $\sqrt{2.25} = 1.5$ per period (50%), and a growth factor exceeding 11,200 over one complete cycle!

The structure of growth rates is displayed in Fig. 9.5 in terms of colour bands, in bundles of 15 (dark blue to yellow) as our software restricts the number of colours. Notably already the first bundle, representing growth rates representing growth from 0.2% to 2.8%, covers about half the area in which we found the periodic tongues whose outlines we can still see in the picture. Yet, above the parabola we still retrieve the unrealistically high growth rates from the original model, further without periodic oscillations, of course.

Conclusion

Above we suggested a business cycle model, consisting of about half of the bits and pieces proposed by Hicks in his classical work. In particular, the "floor" was retained, but the "ceiling" omitted, in concordance with Duesenberry's argument.

As a new element, the floor was tied to actual capital stock through a fixed depreciation factor. Hence, in the process of growth with capital accumulation, the level of the "floor" changes (in the reverse direction to what has been generally assumed), thus allowing increasing amplitude oscillations around the growing trends for income and capital. Further, the secular growth trends are created

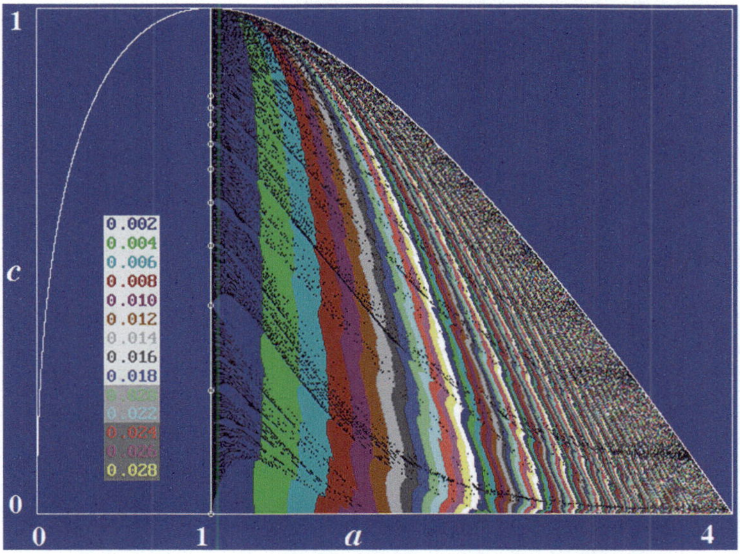

Fig. 9.5 Periodic growth rates for the modified Hicksian trade cycle model with "floor" only range from 0.002 to 0.028, i.e. from 0.2% to 2.8% in the first bundle of the 16 colour range, dark blue to white, (covering about half the interesting range under the parabola). For our exemplifying 23-period cycle with $a = 2.25$ this accrues to about 7.4 over a whole cycle. Compare this to the growth rate in the original model (case with real root) which equals $\sqrt{2.25} = 1.5$ per period (50%)! Over 23 periods this results in a multiplication with a factor about 11,200 over one complete cycle!. Increased realism is here a welcome by-product from the modified model

within the model and need not be introduced in terms of exogenous growing expenditures.[15]

In order to analyze the oscillating growth rates around the rising trends, a transformed system of relative dynamics, in terms of the income growth rate and the capital/income ratio, was proposed. This reduced the system from three to two dimensions, though it also introduced new complexity through transforming linear relations to nonlinear with possibly vanishing denominators.[16]

[15]We may yet want to include autonomous expenditures, because such are not unrealistic, just in order to avoid the absurdity of negative income, but they no longer play any crucial role in providing a growth trend.

[16]In discrete time dynamics this is no big problem. The discontinuity in the pair of hyperbolas $y_t = (a + c) - \frac{a}{y_{t-1}}$ may cause the orbit to very large jumps. However, the relative dynamics would not work in a differential equation system because then the relative variable would indeed become infinite.

9.4 Ceiling Included

9.4.1 Exponential Decay and Half-Life

Above it was suggested that the floor be actually tied to the stock of capital through a depreciation factor r. Then we have $I_t^{\text{floor}} = rK_{t-1}$. This means that we assume a homogeneous capital stock that decays like a radioactive substance. The aggregate stock of capital only wears out in infinite time, but as in the case of radioactivity it may be useful to calculate the half-life n from $(1-r)^n = \frac{1}{2}$, which yields $n = -\frac{\ln 2}{\ln(1-r)}$. For $r = 0.01$, we get $n = 69$, for $r = 0.10$, $n = 7$. Though a vintage approach with capital of various fixed lifetimes would seem more realistic, the above assumption seems to be justifiable at the present level of abstraction.

Given our interpretation of the floor, the investment function we already used in the model with floor only is

$$I_t = \max\left(a\left(Y_{t-1} - Y_{t-2}\right), -rK_{t-1}\right) \tag{9.38}$$

Suppose we use the income formation equation $Y_t = cY_{t-1} + I_t$, without the ceiling, along with $C_t = cY_{t-1}$. Then we have the familiar system

$$Y_t = cY_{t-1} + \max\left(a\left(Y_{t-1} - Y_{t-2}\right), -rK_{t-1}\right) \tag{9.39}$$

and

$$K_t = K_{t-1} + \max\left(a\left(Y_{t-1} - Y_{t-2}\right), -rK_{t-1}\right) \tag{9.40}$$

9.4.2 Tying the Ceiling to Capital Stock

We will now introduce the ceiling Y_t^{ceiling} that appears in (9.2) above, and relate it to the stock of capital. We do this by using the principle of acceleration and the underlying fixed proportions technology once again. Assuming the production function $Y_t = \min\left(\frac{K_{t-1}}{a}, \frac{L_{t-1}}{b}\right)$ it is physically impossible to produce more than $\frac{K_{t-1}}{a}$ or more than $\frac{L_{t-1}}{b}$ in period t.

It is true that Hicks had other production factors such as labour in mind as candidates for the determination of the ceiling, but, to keep the model self-contained, we can even use capital itself as the limiting factor, by putting

$$Y_t^{\text{ceiling}} = \frac{1}{a}K_{t-1}$$

As discussed in the digression above, there is nothing contradictory in this multiple use of the capital stock and capital coefficient. The investors are hunting for the optimal capital stock that corresponds to projected demand, but this does not mean that the process cannot bounce into the ceiling now and then.

What has been said is that we no longer have the simple income formation $Y_t = C_t + I_t$, but use (9.2) with appropriate substitutions for $I_t^{\text{floor}} = rK_{t-1}$ and $Y_t^{\text{ceiling}} = \frac{1}{a}K_{t-1}$. Accordingly

$$Y_t = \min\left(cY_{t-1} + \max\left(a\left(Y_{t-1} - Y_{t-2}\right), -rK_{t-1}\right), \frac{1}{a}K_{t-1}\right) \tag{9.41}$$

replacing (9.39), whereas (9.40) remains as it is.

The new map contains two maximum/minimum clauses, and hence consists of four pieces, each of which is a linear map. See Puu (2007) and Sushko et al. (2010).

The Resulting Map

In order to decide which branch applies we need three conditions. The condition for application of the floor was already stated above, so let us just introduce a symbol for it.

$$R_{t-1} = a\left(Y_{t-1} - Y_{t-2}\right) + rK_{t-1} \tag{9.42}$$

When $R_{t-1} \geq 0$ then the accelerator process works as in the original Samuelson model, when $R_{t-1} < 0$, then the floor is activated. It does not matter whether we take the first as a weak inequality, the second as a strong, or the other way around. When $R_{t-1} = 0$, the accelerator and the floor condition give the same result.

As for the applicability of the ceiling, we check if the sum of consumption cY_{t-1} and investments, accelerator generated as $a\left(Y_{t-1} - Y_{t-2}\right)$, or, fixed at floor level $-rK_{t-1}$, exceed maximum production capacity $\frac{1}{a}K_{t-1}$ or not, so let us define

$$S_{t-1} = cY_{t-1} + a\left(Y_{t-1} - Y_{t-2}\right) - \frac{1}{a}K_{t-1} \tag{9.43}$$

$$T_{t-1} = cY_{t-1} - rK_{t-1} - \frac{1}{a}K_{t-1} \tag{9.44}$$

If $R_{t-1} \geq 0$ and $S_{t-1} \leq 0$, then the classical process works as usual, whereas for $R_{t-1} < 0$ and $S_{t-1} > 0$ we have the exceptional case where both floor and ceiling are applied.

The conditions formulated in terms of (9.42)–(9.44) are not independent, as $S_{t-1} - R_{t-1} = T_{t-1}$. Putting expressions (9.42)–(9.44) equal to zero, we get equations $R_{t-1} = 0$, $S_{t-1} = 0$, and $T_{t-1} = 0$, respectively, which define planes in the $Y_{t-1}, Y_{t-2}, K_{t-1}$-space. However, when $R_{t-1} \geq 0$ holds, then only $S_{t-1} = 0$

is relevant, whereas when $R_{t-1} < 0$ holds, then only $T_{t-1} = 0$ is relevant. Thus we have one plane and two half-planes, the wedge line of the half-planes lying in the plane, as we see from $S_{t-1} - R_{t-1} = T_{t-1}$ stated above. These split the space, not in eight parts as three planes normally would, but in only four, in which we can define the following maps:

$$F_1 : \begin{cases} Y_t = \frac{1}{a} K_{t-1} \\ \\ K_t = K_{t-1} + a\,(Y_{t-1} - Y_{t-1}) \end{cases} \quad R_{t-1} \geq 0, \quad S_{t-1} > 0 \tag{9.45}$$

$$F_2 : \begin{cases} Y_t = cY_{t-1} + a\,(Y_{t-1} - Y_{t-1}) \\ \\ K_t = K_{t-1} + a\,(Y_{t-1} - Y_{t-1}) \end{cases} \quad R_{t-1} \geq 0, \quad S_{t-1} \leq 0 \tag{9.46}$$

$$F_3 : \begin{cases} Y_t = cY_{t-1} - rK_{t-1} \\ \\ K_t = (1-r)\,K_{t-1} \end{cases} \quad R_{t-1} < 0, \quad T_{t-1} \leq 0 \tag{9.47}$$

$$F_4 : \begin{cases} Y_t = \frac{1}{a} K_{t-1} \\ \\ K_t = (1-r)\,K_{t-1} \end{cases} \quad R_{t-1} < 0, \quad T_{t-1} > 0 \tag{9.48}$$

F_1 is the map with ceiling activated, F_3 the one with floor activated, whereas, as noted, F_2 is the regular multiplier—accelerator model, and F_4 the exceptional case where both floor and ceiling are activated.

Note that the origin of the three-dimensional phase space,

$$(Y_{t-1}, Z_{t-1} = Y_{t-2}, K_{t-1}) = (0, 0, 0)$$

is located on all three planes, because it makes $R_{t-1} = S_{t-1} = T_{t-1} = 0$ according to (9.42)–(9.44). As we will see, the origin is a fixed point of all the maps.

Again we have three parameters, $0 < c < 1$, $0 < a$, and $0 < r < 1$.

Equilibrium Fixed Points

To check out the fixed points, let us consider the map branches each in its turn. Note that we used the unusual term "equilibrium fixed points", to distinguish them from "growth/decline fixed points" which we will consider below in the context of the relative system.

Branch F_1

The first map $F1$ obviously has an infinity of fixed points, i.e., any values $Y_t = Y_{t-1} = Y_{t-2} = Y$ and any $K_t = K_{t-1} = K$, such that $Y = \frac{1}{a}K$. However, we easily calculate from (9.42)–(9.43) that $R = rK$ and $S = -\frac{1}{a}(1-c)K$, where we drop the indices for the fixed points. With positive capital we have $R > 0$ and $S < 0$, so the fixed points do not belong to the definition region for F_1. Only when capital is zero, i.e. $K = Y = 0$ does the fixed point belong to the boundary of the region. We did not include the boundary in the definition of branch (9.44), but, as the choice was arbitrary, we can well consider this particular fixed point (out of the infinity) as included in the region.

As for stability, it is easy to derive the factorized characteristic equation: $(\lambda - 1)^2(\lambda + 1) = 0$ with unitary eigenvalues. In order to interpret these facts, note that in (9.45) we can eliminate the income variable, and obtain a difference equation of third order in capital alone: $K_t = K_{t-1} + K_{t-2} - K_{t-3}$, for which we can try the solution $K_t = \lambda^t K_0$. Substituting and rearranging, we obtain $\left(\lambda^3 - \lambda^2 - \lambda + 1\right)K_{t-3} = 0$ as a condition for the attempted solution to fit. Of course, if we are in the fixed point with $K_{t-3} = K = 0$, then any λ would do. But suppose we have a small deviation, in the positive direction, of course, because negative capital does not make any sense, then

$$\left(\lambda^3 - \lambda^2 - \lambda + 1\right) = (\lambda - 1)^2(\lambda + 1) = 0$$

has to hold. This is nothing but the characteristic equation already written. Accordingly, we can have $K_t = (1)^t K_0$, or $K_t = (-1)^t K_0$, The latter case means that capital alternates between positive and negative values every second period, so $\lambda = -1$ is factually irrelevant, whereas the first case means neutral stability. Should capital be perturbed from the zero value, the system immediately moves out from the range of definition for (9.45), and further out of the region that makes any economic sense.

Branch F_2

Next, branch F_2 obviously has infinitely many fixed points $Y_t = Y_{t-1} = Y_{t-2} = 0$; $K_t = K_{t-1} = K$, where K can take any non-negative value. We calculate $R = rK \geq 0$; $S = -\frac{1}{a}K \leq 0$, so, according to (9.46) the fixed points obviously are in the range of definition for this branch. The stability is easily checked from the characteristic equation which factorizes to

$$\left(\lambda^2 - (a+c)\lambda + a\right)(\lambda - 1) = 0$$

The eigenvalues are

$$\lambda_{1,2} = \frac{1}{2}(a+c) \pm \frac{1}{2}\sqrt{(a+c)^2 - 4a}$$

quite as in the original multiplier—accelerator model, and $\lambda_3 = 1$, associated with the neutral stability of the capital stock variable.

Eigenvalues $\lambda_{1,2}$ can be real, if $(a+c)^2 \geq 4a$, or complex conjugates if $(a+c)^2 < 4a$. In either case the fixed point is stable if $a < 1$, unstable if $a > 1$. The boundary case $a = 1$ is exceptional with

$$\lambda_{1,2} = \frac{1}{2}(1+c) \pm i\frac{1}{2}\sqrt{4-(1+c)^2}$$

being complex conjugates (as $0 < c < 1$) of unit modulus ($\lambda_1\lambda_2 = a = 1$). As $F2$ has constant coefficients the map is a contraction or expansion in its entire range of definition depending on whether $a < 1$ or $a > 1$.

Branch F_3

Next, F_3 is the normal floor branch. Its only fixed point is $Y_t = Y_{t-1} = 0$, $K_t = K_{t-1} = K$. We have $R = 0$; $S = 0$, so the point belongs to the boundary of the definition range. This map is two-dimensional and the characteristic equation is $(\lambda - c)(\lambda - (1-r)) = 0$, so both eigenvalues are real, positive and less than unity.

The map is hence a contraction in its entire range of definition.

Branch F_4

Finally, F_4, also two-dimensional, again has $Y_t = Y_{t-1} = 0$, $K_t = K_{t-1} = K$ as its only fixed point, which is included in the boundary of its range. As for stability, one eigenvalue is zero, and the other equals $0 < (1-r) < 1$. From (9.48), the obvious solution for the branch is $K_t = (1-r)^t K_0$ and $Y_t = \frac{1}{a}(1-r)^{t-1} K_0$, so we deal with a contraction, identical with one of the contractions for F_3.

As a conclusion, when $a < 1$, then the composite map is a contraction, whereas, if $a > 1$, we deal with the interesting case where the process jumps between the regions.

We deal with a repulsive fixed point with zero income and arbitrary capital for F_2, combined with an attractive fixed point for $F_3 - F_4$. The process is kept going through jumping between these regions.

9.4.3 The Relative System

As in the case with floor alone we now convert the system (9.45)–(9.48) to a relative system, in order to obtain stationary orbits, using substitutions $y_t = \frac{Y_t}{Y_{t-1}}$ and $k_t = \frac{K_t}{Y_{t-1}}$ for the income growth ratio and the actual capital to income ratio, respectively.

Through the conversion all the branches become second order, which is an additional benefit. In order to define the branches we have to express also branch conditions (9.42)–(9.44) in terms of the new relative variables. To this end we multiply them through by the quotients $\frac{K_{t-1}}{Y_{t-2}^2}$, so obtaining

$$\rho_{t-1} = \frac{K_{t-1}}{Y_{t-2}^2} R_{t-1} = a\left(y_{t-1} - 1\right) k_{t-1} + r k_{t-1}^2 \tag{9.49}$$

$$\sigma_{t-1} = \frac{K_{t-1}}{Y_{t-2}^2} S_{t-1} = (a + c)\, y_{t-1} k_{t-1} - a k_{t-1} - \frac{1}{a} k_{t-1}^2 \tag{9.50}$$

and

$$\zeta_{t-1} = \frac{K_{t-1}}{Y_{t-2}^2} T_{t-1} = c y_{t-1} k_{t-1} - \frac{1}{a} k_{t-1}^2 - r k_{t-1}^2 \tag{9.51}$$

The reason for choosing this complicated factor $\frac{K_{t-1}}{Y_{t-2}^2}$ for the conversion is that if we just divide by Y_{t-1}, which may seem to be an obvious alternative, then, as the variable takes on both positive and negative values, the sense of the inequalities could be reversed, and we would end up at even more branches than we already have. Therefore we need some factor that is guaranteed to be non-negative.

The map now becomes

$$f_1 : \begin{cases} y_t = \frac{1}{a} \frac{k_{t-1}}{y_{t-1}} \\ k_t = \frac{k_{t-1}}{y_{t-1}} + a\left(1 - \frac{1}{y_{t-1}}\right) \end{cases} \quad \rho_{t-1} \geq 0, \sigma_{t-1} > 0 \tag{9.52}$$

$$f_2 : \begin{cases} y_t = c + a\left(1 - \frac{1}{y_{t-1}}\right) \\ k_t = \frac{k_{t-1}}{y_{t-1}} + a\left(1 - \frac{1}{y_{t-1}}\right) \end{cases} \quad \rho_{t-1} \geq \sigma_{t-1} \leq 0 \tag{9.53}$$

$$f_3 : \begin{cases} y_t = c - r\frac{k_{t-1}}{y_{t-1}} \\ k_t = (1-r)\frac{k_{t-1}}{y_{t-1}} \end{cases} \quad \rho_{t-1} < 0, \varsigma_{t-1} \leq 0 \tag{9.54}$$

$$f_4 : \begin{cases} y_t = \frac{1}{a}\frac{k_{t-1}}{y_{t-1}} \\ k_t = (1-r)\frac{k_{t-1}}{y_{t-1}} \end{cases} \quad \rho_{t-1} < 0, \varsigma_{t-1} > 0 \tag{9.55}$$

As we see, there are still four branches. Further, all branches of the system have really become two-dimensional, though they are now nonlinear.

Exemplary Trajectories

To get a first idea of how the system (9.52)–(9.55) works, we present some computer graphics in Figs. 9.6 and 9.7. We calculated trajectories for y_t and Y_t in Fig. 9.6, and for k_t and K_t in Fig. 9.7. As we see, the traces for y_t and k_t are stationary (periodic), whereas Y_t and K_t are growing. The applications of the submaps during one full period for y_t and k_t are indicated by vertical strips in different colour.

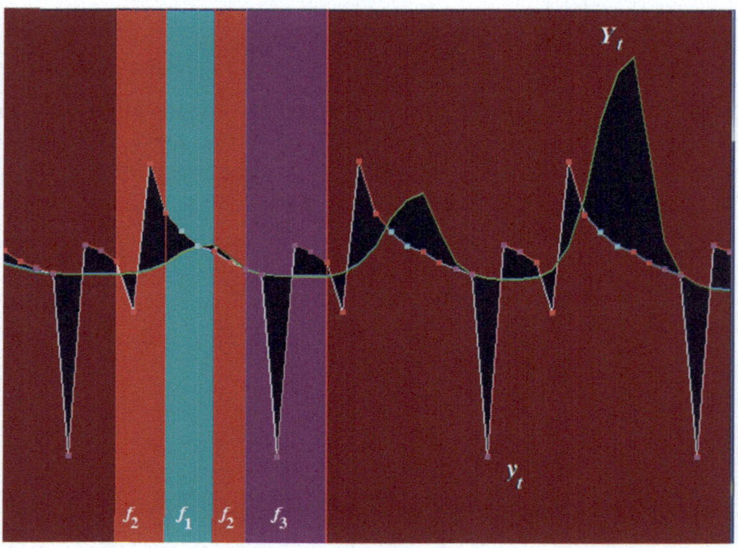

Fig. 9.6 The income trace Y_t oscillating around its growth trend with temporary recessions, and the stationary income growth ratio producing a 13-period growth cycle when the parameters are $a = 3$, $c = 0.35$, and $r = 0.01$. The vertical strips indicate the application of different sub-maps f_i. As we see, in each cycle f_2 is applied twice in a cycle, f_1 and f_3 once each, and f_4 not at all

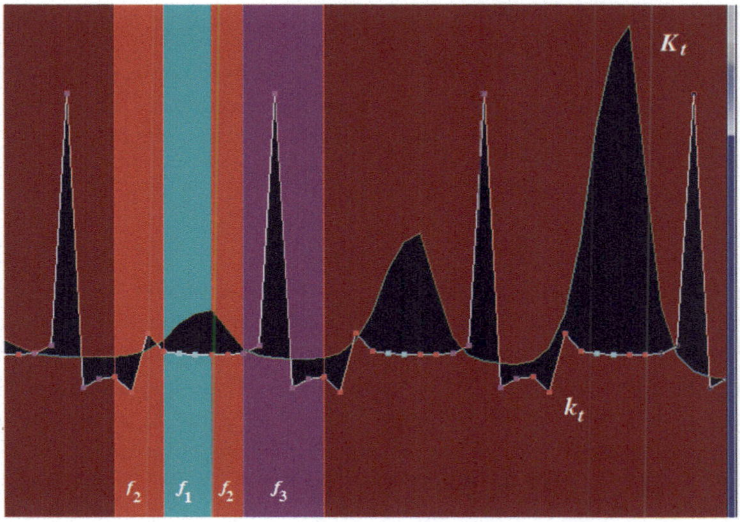

Fig. 9.7 Traces for capital K_t (growing) and for the capital/income ratio k_t (stationary 13-period cycle) with parameters $a = 3$, $c = 0.35$, and $r = 0.01$. The multicoloured vertical strips indicate the application of different maps f_i. As we see, f_2 is applied twice in a cycle, f_1 and f_3 once each, and f_4 not at all

First we have one strip with four applications of f_2, the Samuelson accelerator case. Next come two applications of f_1, the ceiling case, then again two applications of f_2, and, finally, five applications of f_3, the floor case. After that the pattern repeats over and over to produce a 13-period cycle in the income growth ratio. As we see, the combined floor/ceiling map f_4 is never applied in the case represented here.

To obtain these final periodic traces we, of course, scrapped a considerable number of initial iterates, belonging to the transient.

The income trace Y_t in Fig. 9.6 is not stationary, it is growing, and we would not have been able to see any periodicity in it. Growth is with periodic recessions, and we could easily draw a growth trend through the income trace. Likewise, capital K_t is growing with temporary recessions when the floor is activated in the f_3 strips. We should note that the downslope of the capital accumulation curve in the floor regions becomes steeper the higher the accumulated capital stock is, quite as it should be when we relate depreciation, and hence the floor level to capital stock.

The calculations were done by system (9.52)–(9.55), assuming parameters $a = 3$, $c = 0.35$, and $r = 0.01$ to obtain y_t and k_t. Y_t and K_t were then calculated, using the definitions $y_t = \frac{Y_t}{Y_{t-1}}$ and $k_t = \frac{K_t}{Y_{t-1}}$ in reverse, $Y_t = y_t Y_{t-1}$ and $K_t = k_t Y_{t-1}$. Note that we could also have used (9.45)–(9.48) to obtain Y_t and K_t, and then used definitions $y_t = \frac{Y_t}{Y_{t-1}}$ and $k_t = \frac{K_t}{Y_{t-1}}$ as they stand to calculate y_t and k_t.

A note of warning against being too much impressed by such calculated traces should be added. Any such case is but an example, showing one possibility out of a multitude of different ones, which we might not like at all as representations of the qualitative features of business cycles in reality.

Fixed Growth/Decline Points

It is obvious that relative system (9.52)–(9.55) has its fixed points as well. However, now they are fixed points for the growth rate of income, and for the capital/income ratio. So, let us again check them for the different branches, one by one.

Branch f_1

System (9.52) defining the first branch f_1 has two fixed points: $y_t = y_{t-1} = y = \pm 1, k_t = k_{t-1} = k = a$. Referring back to the definitions of the relative coordinates, $y_t = \frac{Y_t}{Y_{t-1}}$ we realize that $y_t = -1$ would mean that $Y_t = Y_{t-1}$ for any t, i.e., that Y_t has some given absolute value, but alternates in sign. This is perfectly possible, but, given $k = a > 0$, we deduce from definition $k_t = \frac{K_t}{Y_{t-1}}$ that $K_t = aY_t = -aY_{t-1}$, i.e., that capital as well oscillates between positive and negative values. This is sheer nonsense, so only $y_t = 1$ makes sense.

However, it is good to know, as we will see, that neither of the fixed points belongs to the definition range for f_1, so that while running the combined system there is no danger to end up at the negative fixed point. Substituting $y = 1, k = a$ in (9.49)–(9.50), we easily get

$$\rho = a^2 r > 0, \sigma = -a(1-r) < 0$$

whereas substituting $y = -1, k = a$, we get

$$\rho = -a^2(1-r) < 0, \sigma = a(2a + c + 1) < 0$$

As $\sigma < 0$, neither of the fixed points belongs to the definition range. There are hence no fixed points in the range for the branch f_1.

Branch f_2

From (9.53) we next find that there are again two fixed points, the two roots of

$$y^2 - (a+c)y + a = 0$$

i.e.,

$$y = \lambda_{1,2} = \frac{1}{2}(a+c) \pm \frac{1}{2}\sqrt{(a+c)^2 - 4a}$$

and, again, $k = a$.

As we see the two roots for y are the same as the eigenvalues $\lambda_{1,2}$ as stated above, which is no coincidence.

As we know, these roots can be real or complex conjugate. In the present context of fixed growth points for the relative system they only exist if they are real.

First, note that both roots $\lambda_{1,2} > 0$ are positive, because the parameters are positive. Second, from $y^2 - (a + c)\, y + a = 0$, we know that $\lambda_1 \lambda_2 = a$, and that $\lambda_1 + \lambda_2 = a + c$. Now, subtract the second equation from the first: $\lambda_1 \lambda_2 - \lambda_1 - \lambda_2 = -c$. Finally, add a unit constant and factorize to obtain: $(\lambda_1 - 1)(\lambda_2 - 1) = 1 - c > 0$. Hence the roots must either both exceed unity or both fall short of unity.

Provided $a > 1$, which is the case of growth, they must both exceed unity. As $\lambda_1, \lambda_2 > 1$, $y > 1$ for both roots.

Let us so check the branch constraints. From (9.49)–(9.50), using $k = a$, we have $\rho = a^2 (y - 1)\, ra^2 > 0$ when $y > 1$. Further, again using $k = a$, we get $\sigma = ((a + c) - a - 1)$. However, from $y^2 - (a - c)\, y + a = 0$, we get $(a - c) - a = y^2$, so using this substitution, finally $\sigma = \left(y^2 - 1\right) > 0$. Hence the fixed growth points $(a > 1)$ of the map f_2 do not belong to its range of its definition.

This conclusion is different from the case with floor alone, where exponential growth was one possibility. The application of the ceiling hence makes simple growth without oscillations impossible. We will see that this is corroborated by numerical experiment below.

Branch f_3

For f_3 we calculate two fixed points from (9.54): $y = c, k = 0$, and

$$y = (1 - r)\,, k = -\frac{1}{a}(1 - r)(1 - c - r)$$

For the first alternative we find $\rho = 0, \varsigma = 0$, so the fixed point is located on the boundary of the definition range for the map. It might be useful to check stability. The eigenvalues are easily calculated, one is 0, the other $\frac{1-r}{c}$, so the fixed point is stable if $c + r > 1$. As $y = c < 1$, the process represents a classical multiplier application (with zero capital), approaching zero income as well.

The alternative fixed point has eigenvalues 0 and $\frac{c}{1-r}$, i.e., the reciprocal of the one for the preceding case. It is stable when $c + r < 1$. Let us just calculate

$$\rho = (1 - r)\left(a\,(1 - c - r) + \frac{1}{r}(1 - c - r)^2\right)$$

which is a bit messy. Given that the point is in the range of definition for the map, we may have $\rho < 0$, so this fixed point may both be admissible and stable, as may the other one.

Branch f_4

Finally, for f_4, we have

$$y = (1 - r), k = a(1 - r)^2$$

To check if it belongs to the definition range calculate

$$\rho = -a^2 r^2 (2 - r)(1 - r)^2 < 0$$

and further

$$\varsigma = -a(1 - r)^2 (ar + (1 - r)(1 - r - c))$$

In order that the fixed point be in the definition range, it is necessary to have $r + c > 1$. The map has two zero eigenvalues. This final branch is where both floor and ceiling apply at once.

9.4.4 Numerical Results

To get some more information, we now have to turn to numerical work. In Fig. 9.8 we show the bifurcation diagram in parameter plane a, c, with fixed $r = 0.025$, where we checked the 100 lowest periodicities. The only issue we can settle by analytical means is where the periodicity tongues set out from the bifurcation line $a = 1$. From map branch F_2 (9.46) we already saw that the eigenvalues are complex conjugates with

$$\lambda_{1,2} = \frac{1}{2}(1 + c) \pm i\frac{1}{2}\sqrt{4 - (1 + c)^2}$$

and unit modulus $\lambda_1 \lambda_2 = a = 1$.

Accordingly we can also write $\lambda_{1,2} = \cos\theta \pm i \sin\theta$ where

$$\cos\theta = \frac{1}{2}(1 + c)$$

and

$$\sin\theta = \frac{1}{2}\sqrt{4 - (1 + c)^2}$$

Using the first, which is simpler, and assuming a periodic solution $\theta = \frac{m}{n}2\pi$, we have

$$\cos\left(\frac{m}{n}2\pi\right) = \frac{1}{2}(1 + c)$$

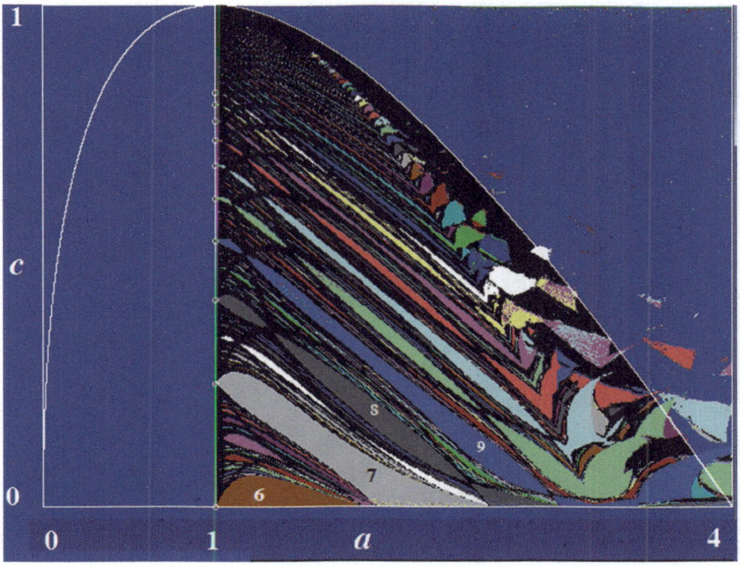

Fig. 9.8 The 100 lowest periodicities in the $a; c$ parameter plane, with $r = 0.025$. Obviously, compared to the case with floor alone, the ceiling introduces much more complexity to the model, maybe even chaotic orbits, which we will check out in the following picture

or

$$c = 2\cos\left(\frac{m}{n}2\pi\right) - 1$$

Putting $m = 1$ and $n = 6, 7, \ldots 15$, we can calculate the dots on the bifurcation line $a = 1$ in Fig. 9.8. (For $n = 2, 3, 4, 5$ we have $c = 2\cos\left(\frac{1}{n}2\pi\right) < 0$, which is unfeasible.) Note that we in fact deal with two different systems. For the maps (9.45)–(9.48), with $a < 1$, we have a stable fixed point in the original model, a node above the parabola, a focus below it. But for the system (9.52)–(9.55) there is just divergence in this part of the bifurcation diagram.

Further, as for the right part of the diagram, with $a > 1$, there are no periodicities in model (9.45)–(9.48), they can only be computed in (9.52)–(9.55). Only in the boundary case, where $a = 1$, do both models make sense at once, and we can also carry out the above calculation. The rest is numerical.

The periodicity tongues thus set out where they did in the model without a ceiling, and it is obvious where the ceiling sets in, along the downsloping line in the right part of the diagram where the shapes become more irregular. Particularly noteworthy is that unlike the case with floor only, the periodicity tongues now protrude through the parabola that separates complex roots from real roots in the original Samuelson model. This corroborates our conclusion that there is no longer any exponential growth without oscillations when the ceiling is added.

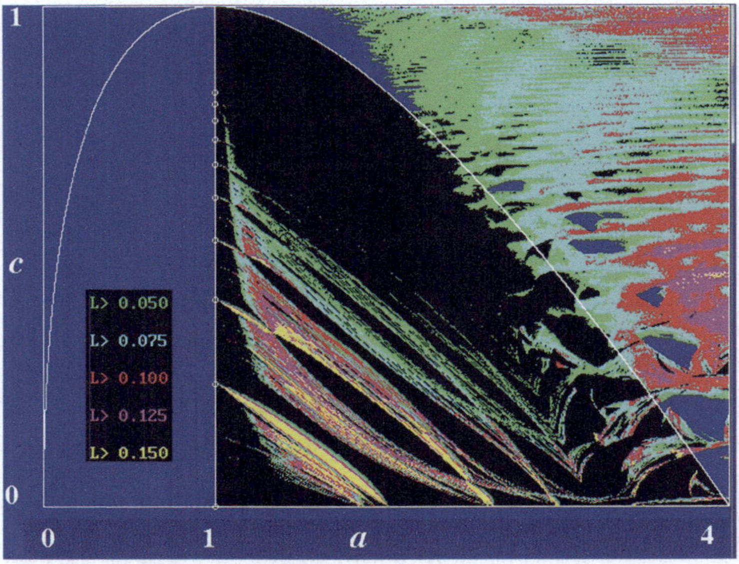

Fig. 9.9 Positive Lyapunov exponent areas in the parameter plane. The colour code is again blue to red representing the value of the highest Lyapunov exponent. As we see there are now extended regions with positive largest Lyapunov exponent, especially outside the parabola. Note that apparently chaotic regions extend through the parabola, thus annihilating pure growth which remained a possibility in the case with floor only

Finally, Fig. 9.9 shows the areas of positive Lyapunov exponents, where the largest exponent is in the interval $(0.05, 0.15)$ as indicated by the different colours (the more reddish the higher).

It can be noted how the coloured areas fit in the blank spaces left by the periodicity tongues in the previous figure.

Note that high Lyapunov exponent areas, quite as the periodic tongues, set out just at the bifurcation points, something worth to explore more.

References

Allen RGD (1956) Mathematical economics. Macmillan, London

Duesenberry J (1950) Hicks on the trade cycle. Q J Econ 64:464–476

Gallegati M, Gardini L, Puu T, Sushko I (2003) Hicks's trade cycle revisited: cycles and bifurcations. Math Comput Simul 63:505–527

Gandolfo G (1985) Economic dynamics: methods and models. North-Holland, Amsterdam

Goodwin RM (1951) The nonlinear accelerator and the persistence of business cycles. Econometrica 19:1–17

Hicks JR (1950) A contribution to the theory of the trade cycle. Oxford University Press, Oxford

Hommes CH (1991) Chaotic dynamics in economic models. Wolters-Noodhoff, Groningen

Palander T (1953) On the concepts and methods of the "Stockholm School". International economic papers no. 3, Macmillan

Puu T (1963) A graphical solution to second order homogeneous difference equations. Oxf Econ Pap 15:53–58

Puu T (1987) Complex dynamics in continuous models of the business cycle. In: Lecture notes in economics and mathematical systems, vol 293. Springer, Berlin, pp 227–259. ISBN 3-540-18183-0

Puu T (2007) The Hicksian trade cycle with floor and ceiling dependent on capital stock. J Econ Dyn Control 31:575–592

Puu T, Gardini L, Sushko I (2005) A multiplier-accelerator model with floor determined by capital stock. J Econ Behav Organ 56:331–348

Rau N (1974) Trade cycle: theory and evidence. Macmillan, London

Samuelson PA (1939) Interactions between the multiplier analysis and the principle of acceleration. Rev Econ Stat 21:75–78

Sushko I, Gardini L, Puu T (2004) Tongues of periodicity in a family of two-dimensional maps of real Möbius type. Chaos Solitons Fractals 21:403–412

Sushko I, Gardini L, Puu T (2010) Regular and chaotic growth cycles in a Neo-Hicksian floor/ceiling model. J Econ Behav Organ 75:77–99

von Haberler G (1937) Prosperity and depression. Harvard University Press, Cambridge

Part III
Model Stubs

Chapter 10
Introduction to the Stubs

In this last part we suggest a number of research "stubs"; model ideas, formulated but not analyzed by the present author.

The analysis is left to the interested reader. Some stubs may be appropriate for an article, some might even be sufficient for a thesis, especially if obvious variations on the themes are included.

Of course, it is easy to come up with new model ideas. However, everybody who tried hands on this, has experienced many stumbling blocks. Such models may just be complex and messy and thus not yield any precise results; they may further not be analyzable by known methods.

The common feature of these stubs to follow is that the author knows that they yield to analysis and may result in precise and nontrivial results. In fact the author carried the analysis further than appears in the text, but stopped short before that point as he did not wish to deprive the reader of the pleasure of finding out something more.

What is particular to economics is that any good model results from some imagined "model reality". Hence different such models emerge from different model realities created to focus different issues, and they cannot just be mixed in some new model.

Above we showed a, hopefully credible, way to blend the Cournot and Stackelberg models.

However, we also criticized recent confused attempts to mix Cournot and Bertrand oligopolies. This mix does not work because the original model worlds are contradictory. Below it is suggested how the issue can be rectified using multiple markets and discriminatory pricing.

As a rule model blends are difficult and call for entering the basics through creating some new model world which is credible in itself. The blending can never be a mechanical process because economics does not deal with established laws of nature.

T. Puu, *Disequilibrium Economics*, https://doi.org/10.1007/978-3-319-74415-5_10

Notably such problems occur also in disciplines that do deal with laws of nature such as physics, as exemplified through blending relativity and quantum mechanics in the case of black holes. Some researchers claim that the proposed solution, superstring theory, can never render any independent test implications. The solution, though intellectually much more sophisticated than any economics, simply moved such way towards pure abstraction so as to approach the model worlds of economics.

Chapter 11
CAUDEX PRIMUS

Abstract Recently the present author read a few publications which tried to combine Cournot and Bertrand duopoly. The approach, using linear demand functions with appropriate signs for cross derivatives, seemed basically mistaken. **First**, linear functions never work in global dynamics. **Second**, linear demand functions are not consistent with utility maximization because they do not fulfil budget constraints. They can locally approximate proper demand functions, but, again, do not work globally. **Third**, the whole idea is contradictory, because a commodity is either homogenous, or it is not, "**tertium non datur**". It depends on how the consumers regard the case, and is nothing for the duopolists to decide. This made the present author think how one might bring some logic in this. Actually, price discrimination is standard in economic theory, and often applied in practice, so one could consider a case with two different groups of consumers, one that regards the oligopolists' produce as equivalent, and another which regards their supply as different but close substitutes. Then both suppliers sell to both groups and apply price discrimination. In this way the competitors can be Cournot duopolists in one market and Bertrand duopolists in another.

There has been an apparent misunderstanding concerning the use of price or quantity as "action parameter" as it was once called. In the Cournot market, of course, price cannot be used as there is just one price for the as homogenous conceived commodity and this depends on both supplies. However, for the Bertrand markets for the brands, price or quantity can be used for optimization! It makes no difference as the markets are monopoly markets and there is only one degree of freedom once the demand function is given. It is even convenient to prefer quantity as "action parameter" even here: Traditional microeconomics has been set up this way as displayed through Amoroso's Formula used in even pure monopoly.

To think the difference between the Cournot and Bertrand cases lies in which variable to use for optimization reveals a misunderstanding of elementary microeconomics.

T. Puu, *Disequilibrium Economics*, https://doi.org/10.1007/978-3-319-74415-5_11

11.1 A Formal Model

11.1.1 Demand

So assume the two competitors supply $q_1 + Q_1$ and $q_2 + Q_2$, the quantities in lower case in the markets for their own proper brand at prices p_1 and p_2 respectively, and those in upper case in a common market at the price P.

For the demand functions it is simple and convenient to choose the isoelastic

$$P = \frac{1}{Q_1 + Q_2}$$

again for the common market. As for the discriminating markets the choice is more difficult due to the shortage of useful global demand functions for close substitutes.

We choose the Lancaster case again, though we will introduce it in a different way from Chap. 5 to see other facets. Recall that the properties that entered utility were

$$x_1 = a_{11}q_1 + a_{12}q_2$$
$$x_2 = a_{21}q_1 + a_{22}q_2$$

and assume a Cobb-Douglas utility $U = x_1 x_2$. As we deal with ordinal utility we can delete any usual square root or any other monotonic transformation. Hence, $U = (a_{11}q_1 + a_{12}q_2)(a_{21}q_1 + a_{22}q_2)$.

Rescaling, through division by $a_{12}a_{21}$, further using $b_1 = \frac{a_{11}}{a_{12}}$ and $b_2 = \frac{a_{22}}{a_{21}}$ as introduced in Chap. 5, we get

$$U = b_1 q_1^2 + (1 + b_1 b_2)\, q_1 q_2 + b_2 q_2^2$$

Further, dividing through by half the coefficient of the product term, and using the approximation amplitudes also defined in Chap. 5 $A_1 = \frac{2b_2}{1+b_1 b_2}$ and $A_2 = \frac{2b_1}{1+b_1 b_2}$

$$U = A_2 q_1^2 + 2q_1 q_2 + A_1 q_2^2$$

Notably the utility function implicit with the use of Lancaster's transformation is quadratic. As all the design parameters are positive, it has a unique absolute maximum, inconveniently located at the origin. But this can easily be remedied through replacing $(q_1 - 1) \rightarrow q_1$ and $(q_2 - 1) \rightarrow q_2$. Again this is licit with ordinal utility. In this way we get a positive maximum in the positive quadrant.

But still, there is one maximum, now at $q_1 = q_2 = 1$, and for

$$(q_2 - 1) > A_2 (q_1 - 1)$$

or

$$(q_2 - 1) > A_1 (q_1 - 1)$$

we cannot use optimization with a linear budget constraint because the second order conditions do not hold so any inner solutions return minima in stead of maxima.

These problems with reasonable solutions in restricted regions only relate to the discussion of degeneracy for the determinant of the Lancasterian system. Restating utility with the substitutions we maximize

$$U = A_2 (q_1 - 1)^2 + 2 (q_1 - 1) (q_2 - 1) + A_1 (q_2 - 1)^2$$

subject to the usual budget constraint

$$p_1 q_1 + p_2 q_2 = 1$$

The resulting demand functions are easily obtained as

$$q_1 = \frac{A_1 p_1 - p_2}{A_1 p_2^2 - 2 p_1 p_2 + A_2 p_2^2}$$

$$q_2 = \frac{A_2 p_2 - p_1}{A_1 p_2^2 - 2 p_1 p_2 + A_2 p_2^2}$$

In order to provide linear demand functions with some underpinning, one often sees the claim that a quadratic utility function would give rise to such, but, as we see, this is simply not true, the demand functions derived from quadratic utility are far more complicated than that.

In order to conform to the Amoroso format we invert the system to obtain the inverse demand

$$p_1 = \frac{A_2 q_1 + q_2}{A_1 q_2^2 + 2 q_1 q_2 + A_2 q_1^2}$$

$$p_2 = \frac{A_1 q_2 + q_1}{A_1 q_2^2 + 2 q_1 q_2 + A_2 q_1^2}$$

11.1.2 Cost and Profit

Assuming constant marginal costs c_1, c_2 we have all bits and pieces for composing the profits of the firms

$$V_1 = p_1 q_1 + P Q_1 - c_1 (q_1 + Q_1)$$

$$V_2 = p_2 q_2 + P Q_2 - c_2 (q_2 + Q_2)$$

or, substituting for p_1, p_2 and P,

$$V_1 = \frac{A_2 q_1^2 + q_1 q_2}{A_1 q_2^2 + 2q_1 q_2 + A_2 q_1^2} + \frac{Q_1}{Q_1 + Q_2} - c_1 (q_1 + Q_1)$$

$$V_2 = \frac{A_1 q_1^2 + q_1 q_2}{A_1 q_2^2 + 2q_1 q_2 + A_2 q_1^2} + \frac{Q_2}{Q_1 + Q_2} - c_2 (q_2 + Q_2)$$

11.1.3 Reaction Functions

To obtain the reaction functions equate the derivatives to zero, $\frac{\partial V_1}{\partial q_1} = 0$, $\frac{\partial V_1}{\partial Q_1} = 0$, $\frac{\partial V_2}{\partial q_2} = 0$, $\frac{\partial V_2}{\partial Q_2} = 0$.

Then, for the Cournot market (upper case variables) we can as in Chap. 3, easily solve for the reaction functions

$$Q_1' = \sqrt{\frac{Q_2}{c_1}} - Q_2$$

$$Q_2' = \sqrt{\frac{Q_1}{c_2}} - Q_1$$

whereas for the difficult problem of the Bertrand markets we can use the approximations

$$q_1' = \sqrt{A_1}\sqrt{\frac{q_2}{c_1}} - A_1 q_2$$

$$q_2' = \sqrt{A_2}\sqrt{\frac{q_1}{c_2}} - A_2 q_1$$

The two pairs of iterations, however, are independent, and we already studied each separately.

11.2 Capacity Limits

As we see, with constant marginal costs, any dynamic processes according to Cournot and Bertrand become independent processes combining such that we studied already. To make them dependent and produce new phenomena, it might be interesting to assume capacity limits on the cost side.

Hence, take the convenient CES production function used in Chap. 4, $q = \frac{kl}{k+l}$. Taking capital k as fixed over an investment period, we can solve for the input of labour $l = \frac{kq}{k-q}$. This function form is convenient as the fixed capital stock provides a capacity limit, so that labour, and accordingly labour cost $wl = w\frac{kq}{k-q}$ goes to infinity when $q \to k$.

Replacing the cost functions $C = c\,(q + Q)$ for the discriminating duopolists by $C = rk + w\frac{k(q+Q)}{k-(q+Q)}$, we get marginal cost $MC = w\frac{k^2}{(k-q-Q)^2}$. Accordingly, the cost side introduces a dependence between the systems for the Cournot market and for the Bertrand brand markets that emerge from discrimination.

The profits we want to maximize are now

$$V_1 = \frac{A_2 q_1^2 + q_1 q_2}{A_1 q_2^2 + 2q_1 q_2 + A_2 q_1^2} + \frac{Q_1}{Q_1 + Q_2} - rk_1 - w\frac{k_1\,(q_1 + Q_1)}{k_1 - (q_1 + Q_1)}$$

$$V_2 = \frac{A_1 q_1^2 + q_1 q_2}{A_1 q_2^2 + 2q_1 q_2 + A_2 q_1^2} + \frac{Q_2}{Q_1 + Q_2} - rk_2 - w\frac{k_2\,(q_2 + Q_2)}{k_2 - (q_2 + Q_2)}$$

The expressions

$$\frac{\partial V_1}{\partial Q_1} = \frac{Q_2}{(Q_1 + Q_2)^2} - w\frac{k_1^2}{(k_1 - (q_1 + Q_1))^2} = 0$$

$$\frac{\partial V_2}{\partial Q_2} = \frac{Q_1}{(Q_1 + Q_2)^2} - w\frac{k_2^2}{(k_2 - (q_2 + Q_2))^2} = 0$$

result in nice reaction functions we can easily solve for, but $\frac{\partial V_1}{\partial q_1} = 0$ and $\frac{\partial V_2}{\partial q_2} = 0$ are bound to lead to quite a lot of juggling. As they could not be solved for reaction functions when marginal costs were constant, they are even more unlikely to be solved with more complicated cost functions.

Perhaps again some approximation could work. To this end we need to calculate reaction functions numerically to find the general shape, and to fit the approximations numerically quite as in Chap. 5.

Reference

Tremblay CH, Tremblay VO (2011) The Cournot-Bertrand model and the degree of product differentiation. Econ Lett 111:233–235

Chapter 12
CAUDEX SECUNDUS

Abstract The enigmatic Hotelling duopoly that we discussed in Chap. 7 has never been set in the proper geographical space of two dimensions. Already the case of two sellers on a line interval may be considered technically tricky, especially if we want elastic demand, which we do. The author has thought about this perspective over a long period and also seen many attempts by others, essentially still one dimensional with a thick strip replacing the thin line interval. But this is not what we want. Frankly speaking, it is not even obvious what the Hotelling problem in two dimensions is. Is it three competitors in an equilateral triangle, or what? Though, it seems that not all possible approaches are unmanageable. The main problem is the Euclidean distance which makes integrations difficult (even a market boundary is defined by a quartic equation). And it is not even realistic! If we consider a Manhattan metric in stead, then market boundaries become straight lines in only horizontal, vertical, or diagonal directions, and this simplifies everything. The standard model with inelastic demand even becomes trivial. Anyhow, there are innumerable interesting issues to explore.

12.1 Hotelling in 2D

The Hotelling spatial duopoly was never transferred to the two dimensions of natural geographical space. Also recall that from intuition the Euclidean metric space is bound to produce messy computations, as we know from Launhardt's study in the nineteenth century that market boundaries become defined by quartic equations.

The Manhattan case promises to be much easier to deal with, though still quite challenging. It may be advisable to start with a region in the shape of a tilted square with corners [1, 0], [0, 1], [−1, 0], and [0, −1]. In conformity, the roads could be horizontal and vertical so that the metric is $d = |x| + |y|$. A little reflection discloses that all market boundaries become parallel to the sides of the basic square—either sides of this square or interior to it, unless two such squares for market areas of competing firms overlap. If so a market boundary which is a horizontal or vertical line segment separates the competing duopoly markets. See Figs. 12.1 and 12.2.

T. Puu, *Disequilibrium Economics*, https://doi.org/10.1007/978-3-319-74415-5_12

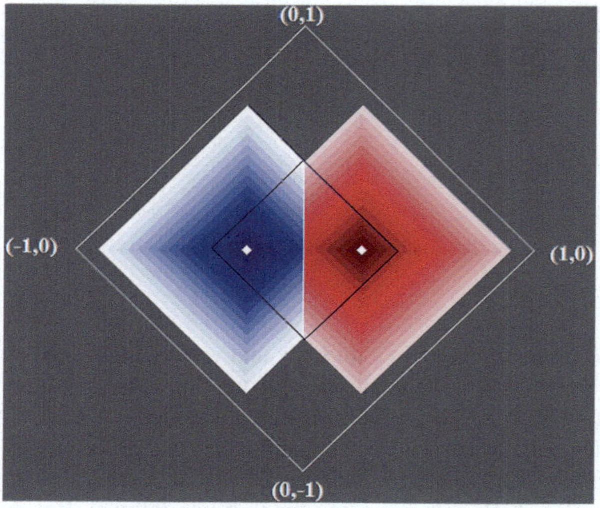

Fig. 12.1 The Hotelling problem in two dimensions. The region is the tilted square of the picture with underlying dense Manhattan metric (West-East and South-West directions of roads). The two competing firms are here symmetrically located at the two bright spots. Their potential market areas (blue and red squares) have a considerable overlap, which is divided by a vertical straight line. Equal price contours are indicated by shading. We could imagine the picture to display what we see if we look down from zenith on a pair of intersecting (square) Launhardt funnels. The blue and red squares are supposed to be delimited where the local delivered price (mill price plus transport cost) reach the maximum that consumers are willing to pay. This leaves a gray area where the commodity is not supplied by any firm as it would become too expensive to buy. We also see the black outline of the intersection of potential market areas which is split between the firms. This shows how demand for each competitor can simplest be calculated through demand for the entire square market minus half the demand for the overlap (again square) in conformity to the rest. We can imagine that, using the Manhattan metric, the 2D case is only slightly more complicated than the 1D case studied in innumerable contributions

In summary, any market is bounded by horizontal/vertical or $\pm 45°$ segments—at most of an octagonal shape. These shapes conform with the metric assumed, which facilitates integrations.

Suppose we have a firm located at $[x_1, y_1]$ in our square with a radius of r_1, for simplicity totally interior to the region square. The market area is then

$$A_1 = \int_{x_1-r_1}^{x_1+r_1} \int_{y_1-r_1}^{y_1+r_1} dydx = 4r_1^2$$

Likewise, again having a linear demand function

$$q_1 = a - b\left(p_1 + k\left(|x - x_1|\right) + |y - y_1|\right)$$

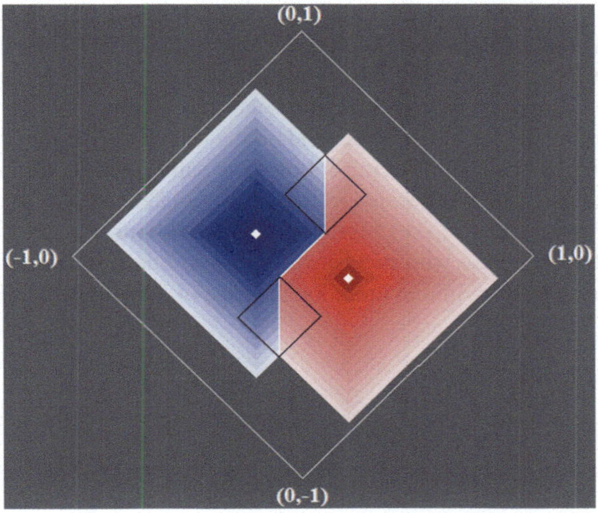

Fig. 12.2 Market areas and delivered prices when the locations of the competitors are not both on the diagonal. As we see, the market boundary then takes a slightly more complicated shape in the form of a train of three segments. But the slopes of these are still in the directions of either the roads (horizontal/vertical) or of the region boundary. For clever integrations we now have two tilted squares indicated in black half of which can be subtracted

we get market demand

$$Q_1 = \int_{x_1-r_1}^{x_1+r_1} \int_{y_1-r_1}^{y_1+r_1} (a - b\,(p_1 + k\,(|x - x_1|) + |y - y_1|))\,dydx$$

which becomes

$$Q_1 = 4(a - b(p_1 + k{\cdot}r_1))r_1^2$$

In case the market area of firm 1 is cut off by the boundary of the entire square the shape becomes a bit asymmetric and the formulas slightly more complex.

Likewise, if there is a competing firm, the boundary segment between the competing areas becomes a horizontal or vertical line.[1]

[1] *Hint*: Though there are many ways to calculate area or market demand for the remaining square truncated by one corner, the easiest calculation is through considering that the truncated corner is half of one of our basic tilted square shapes, and subtract half that lost area or demand. For this virtual area we then have to calculate the centre, the radius, and a virtual supply price.

It may be easiest to start dealing with two firms that only locate along the horizontal diagonal of the square region as shown in Fig. 12.1. Thereafter one can consider other locations and add more competitors Having the general expressions for the total demand of each competitor, one can easily solve for the optimal locations, and investigate how they depend on the transport cost rate. See Fig. 12.2.

Then finding optimal mill pricing follows the same lines as in the one dimensional case discussed. It is advisable to make numerical examples at the derivation of each new formula, and to sketch graphs of the problems at hand. One can study the original Hotelling case with inelastic demand, to find if the paradox extends to two dimensions, or deal with the more logical case of a linear demand function.

Chapter 13
CAUDEX TERTIUS

Abstract The business cycle model proposed in Chap. 9 calls for one obvious generalization; from a closed and isolated economy to open economies connected by international/interregional trade. The Metzler 1950 model with constant import propensities provides an obvious frame for such generalization. A natural starting point is a case with two equal economies. Our business cycle model would provide for growth trends and growth rate cycles in each region, which would now be modified trough external influences from the other economy. It is well known for coupled oscillators that they have two basic modes of operation: in phase, or in opposite phase, and that any outcome can be considered as a mixture of these. An obvious agenda is to check this out for the business cycle model. A more difficult problem is described at the end, i.e., how to model the interaction between economies of different size.

13.1 The Open Economy

The business cycle models discussed were confined to one closed economy. However, we can easily generalize to open economies linked through export and import. So, suppose we confine the discussion to the simplest case—two regions. We now identify the regions through suffices 1, 2, not to be confused with powers.

The accounting identities then read

$$Y_t^1 = C_t^1 + I_t^1 + X_t^1 - M_t^1$$

$$Y_t^2 = C_t^2 + I_t^2 + X_t^2 - M_t^2$$

where X_t^1, X_t^2 denote exports and M_t^1, M_t^2 imports. As a matter of course $X_t^1 = M_t^2$ and $M_t^1 = X_t^2$. Thus, if we subtract these equations, the export/import terms cancel, and we are back to the well-known model for a closed economy, though stated for the difference $Y_t^1 - Y_t^2$.

T. Puu, *Disequilibrium Economics*, https://doi.org/10.1007/978-3-319-74415-5_13

The simplest assumptions concerning parameters $0 < c < 1, 0 < a, 0 < r < 1$, are that they are equal for both regions, so we do not give them any suffices. A common assumption is that imports, like consumption of domestic goods are proportional to income, and that exports are proportional to income abroad. We give this proportionality coefficient the symbol m, again the same for both regions. Obviously $0 < m$, and further $m < c < 1$. Otherwise more than the entire consumption would be imported, which is possible but very unlikely. For one thing, also investment goods may have to be imported, for another such extreme imbalance in foreign trade would not be sustainable in the long run. Thus, $X_t^1 = M_t^2 = mY_{t-2}^2$ and $M_t^1 = X_t^2 = mY_{t-2}^1$.

Substituting along with the usual determinants for consumption, $C_t^1 = cY_{t-1}^1$, $C_t^2 = cY_{t-1}^2$ the equations become

$$Y_t^1 = cY_{t-1}^1 + I_t^1 + mY_{t-1}^2 - mY_{t-1}^1$$

$$Y_t^2 = cY_{t-1}^2 + I_t^2 + mY_{t-1}^1 - mY_{t-1}^2$$

As for investment, the expressions (for the case without ceiling) are not affected:

$$I_t^1 = \max\left(a\left(Y_{t-1}^1 - Y_{t-2}^1\right), -rK_{t-1}^1\right)$$

$$I_t^2 = \max\left(a\left(Y_{t-1}^2 - Y_{t-2}^2\right), -rK_{t-1}^2\right)$$

Neither are the capital accumulation equations.

So, stating the entire system

$$\begin{cases} Y_t^1 = cY_{t-1}^1 + \max\left(a\left(Y_{t-1}^1 - Y_{t-2}^1\right), -rK_{t-1}^1\right) + mY_{t-1}^2 - mY_{t-1}^1 \\ Y_t^2 = cY_{t-1}^2 + \max\left(a\left(Y_{t-1}^2 - Y_{t-2}^2\right), -rK_{t-1}^2\right) + mY_{t-1}^1 - mY_{t-1}^2 \\ K_t^1 = K_{t-1}^1 + \max\left(a\left(Y_{t-1}^1 - Y_{t-2}^1\right), -rK_{t-1}^1\right) \\ K_t^2 = K_{t-1}^2 + \max\left(a\left(Y_{t-1}^2 - Y_{t-2}^2\right), -rK_{t-1}^2\right) \end{cases}$$

It is obviously six-dimensional, if we recall the twice lagged income variables in the accelerators. Obviously it also has four regions of application, depending on whether one or both regions are in phases of accelerator generated growth or of stagnation. The branch conditions would be

$$R_{t-1}^1 = a\left(Y_{t-1}^1 - Y_{t-2}^1\right) + rK_{t-1}^1$$

$$R_{t-1}^2 = a\left(Y_{t-1}^2 - Y_{t-2}^2\right) + rK_{t-1}^2$$

We accordingly have four maps

$$F_1: \begin{cases} Y_t^1 = cY_{t-1}^1 + a\left(Y_{t-1}^1 - Y_{t-2}^1\right) - m\left(Y_{t-1}^1 - Y_{t-1}^2\right) \\ Y_t^2 = cY_{t-1}^2 + a\left(Y_{t-1}^2 - Y_{t-2}^2\right) - m\left(Y_{t-1}^2 - Y_{t-1}^1\right) \\ \\ K_t^1 = K_{t-1}^1 + a\left(Y_{t-1}^1 - Y_{t-2}^1\right) \\ K_t^1 = K_{t-1}^1 + a\left(Y_{t-1}^1 - Y_{t-2}^1\right) \end{cases}$$

applicable when $R_{t-1}^1 \geq, R_{t-1}^2 \geq 0$ holds

$$F_2: \begin{cases} Y_t^1 = cY_{t-1}^1 + a\left(Y_{t-1}^1 - Y_{t-2}^1\right) - m\left(Y_{t-1}^1 - Y_{t-1}^2\right) \\ Y_t^2 = cY_{t-1}^2 - rK_{t-1}^2 - m\left(Y_{t-1}^2 - Y_{t-1}^1\right) \\ \\ K_t^1 = K_{t-1}^1 + a\left(Y_{t-1}^1 - Y_{t-2}^1\right) \\ K_t^2 = (1-r)\,K_{t-1}^2 \end{cases}$$

in case $R_{t-1}^1 \geq 0$, $R_{t-1}^2 < 0$,

$$F_3: \begin{cases} Y_t^1 = cY_{t-1}^1 - rK_{t-1}^1 - m\left(Y_{t-1}^1 - Y_{t-1}^2\right) \\ Y_t^2 = cY_{t-1}^2 + a\left(Y_{t-1}^2 - Y_{t-2}^2\right) - m\left(Y_{t-1}^2 - Y_{t-1}^1\right) \\ \\ K_t^1 = (1-r)\,K_{t-1}^1 \\ K_t^1 = K_{t-1}^1 + a\left(Y_{t-1}^2 - Y_{t-2}^2\right) \end{cases}$$

for $R_{t-1}^1 < 0$, $R_{t-1}^2 \geq 0$, and

$$F_4: \begin{cases} Y_t^1 = cY_{t-1}^1 - rK_{t-1}^1 - m\left(Y_{t-1}^1 - Y_{t-1}^2\right) \\ Y_t^2 = cY_{t-1}^2 - rK_{t-1}^2 - m\left(Y_{t-1}^2 - Y_{t-1}^1\right) \\ \\ K_t^1 = (1-r)\,K_{t-1}^1 \\ K_t^2 = (1-r)\,K_{t-1}^2 \end{cases}$$

for $R_{t-1}^1 < 0$, $R_{t-1}^2 < 0$.

There are several interesting issues to explore. We can check out the equilibrium points and their stability. As for the dynamic behaviour of the model it is good to remember that the regions are identical in terms of parameters, and that coupled equal oscillators use to have two basic modes of oscillation—in the same phase, and in opposite phase. With the accelerator branches this obviously holds, and we can check this out by forming the sums and differences of regional incomes.

However, the traces will still be growing. To reduce them to stationary periodic growth rates we can again use the relative variables, now for both regions

$y_t^1, y_t^2, k_t^1, k_t^2$. We also get a new variable $\frac{Y_t^2}{Y_t^1}$ and its reciprocal $\frac{Y_t^1}{Y_t^2}$ from the export/import terms, even if it may have infinite discontinuity points in the nonlinear model. There is no doubt that these ratios too become stationary, though we need a new variable and a clever updating rule for it in the maps.

Converting the region definitions so that they hold for the relative variables also takes some caution as we want to avoid dividing inequalities by any negative expression.

Once the maps are defined, computer work might show interesting things, for instance concerning synchronization or its opposite.

13.2 Addendum: Small and Big Economies

As a further perspective we may consider the following problem: Through the model setup the coupled economies automatically become equal in size. Now in trade there obviously are big economies and small economies, but the question is how we introduce small and big in the model. The situation is quite like that in Chap. 4, where we in order to consider Cournot's agenda for competition attempted to model small and big firms. There the solution was in terms of capital stocks that provided capacity limits.

We obviously need something similar here. It is always a good idea to start thinking about what is more or less implicit in the model already when starting to generalize. Referring back to the Leontief technology and the original Hicksian reasoning behind the floor, labour force would seem to provide such a possibility. A big economy would have a large population and hence labour force, providing for a high ceiling.[1]

We would then have to reintroduce the original ceiling related to the labour force of the economy, and find a way how to handle this in the relative dynamics setting. Further it may show that a high floor as a passive constraint is not enough to model different sizes of economies, so that we would need some fundamental modifications of the model. Finally, we would have to tune the import propensities; a big economy trades with many small economies, so assuming equal propensities would lead to imbalances in trade which are not sustainable.

These issues are interesting but quite demanding, calling for creativity on a different level than the more or less finished stubs.

Reference

Metzler LA (1950) A multiple- region theory of income and trade. Econometrica 18:329–354

[1]Another alternative would be autonomous public spending which, of course, is more substantial in a big economy.

Chapter 14
CAUDEX QUARTUS

Abstract In Chap. 6 we integrated Stackelberg action as a choice alternative in the dynamic Cournot process. It turned out that such a strategy might be more profitable than either Cournot equilibrium or Stackelberg leadership. The question was also raised why Stackelberg never considered dynamic action in just his own model: If the follower indeed follows the proper Cournot best reply at every move the leader chooses to take, why does the leader not design a changing supply policy over time so as to yield maximum total profit over a longer period? In this stub we set up the leader's dynamic production planning formally. We immediately see that the usual stationary Stackelberg conditions do **not** satisfy the dynamic optimum conditions. These may be difficult to derive explicitly—with finite planning horizon the problem gets a strategic planning character. There, however, remains a lot of interesting issues for numerical study.

14.1 Dynamic Stackelberg

Indeed, the present author checked out the case for Stackelberg's original model with exponential demand, though skipping the square root and using a nonzero production cost. The consequence of the last assumption was that it was not possible to solve for the reaction functions (Puu 2010). However, it was possible to solve for their inverses. It was an interesting task to work with inverse functions giving the competitor's supply that corresponds to each possible best reply of one's own, though a bit awkward. Anyhow the result was that a leadership varying over time was a better choice than keeping to a constant. It may be interesting to see how things work out using our isoelastic demand function instead, which was never done. So, even if there are only two firms, the leader and the follower, we choose the symbols q_t and Q_t to avoid suffices, and similarly the costs c and C. Suppose the upper case firm is the follower and the lower case firm the leader.

Then we always have

$$Q_t = \sqrt{\frac{q_{t-1}}{C}} - q_{t-1}$$

T. Puu, *Disequilibrium Economics*, https://doi.org/10.1007/978-3-319-74415-5_14

The profits of the leader over n periods are

$$V = \sum_{t=1}^{t=n} \left(\frac{q_t}{q_t + Q_t} - cq_t \right)$$

so, substituting for the reactions of the follower,

$$V = \sum_{t=1}^{t=n} \left(\frac{q_t}{q_t + \sqrt{\frac{q_{t-1}}{C}} - q_{t-1}} - cq_t \right)$$

Note that V only depends on the q_t, as Q_t have been substituted out by passive best reply actions—with one exception Q_1. There is nothing before period 1, so the follower could not react to anything, so the first entry remains a kind of initial value.

Now note that if we want to optimize with respect to some special entry then it is contained in two terms in the sum, the t:th and the $t-1$:st, so

$$V = \ldots + \frac{q_t}{q_t + \sqrt{\frac{q_{t-1}}{C}} - q_{t-1}} + \frac{q_{t+1}}{q_{t+1} + \sqrt{\frac{q_t}{C}} - q_t} \ldots - cq_t - cq_{t+1}$$

Accordingly, after some simplification

$$\frac{\partial V}{\partial q_t} = \frac{\left(\sqrt{\frac{1}{bq_{t-1}}} - 1\right)}{q_{t-1}\left(\frac{q_t}{q_{t-1}} + \sqrt{\frac{1}{bq_{t-1}}} - 1\right)^2} - \frac{\frac{q_{t+1}}{q_t}\left(\frac{1}{2}\sqrt{\frac{1}{bq_t}} - 1\right)}{x_t\left(\frac{q_{t+1}}{q_t} + \sqrt{\frac{1}{bq_t}} - 1\right)^2} - c$$

which we put equal to zero. We see that, in principle, the resulting equation lets us determine q_t when q_{t-1} and q_{t+1} are known. This is not a normal iteration, rather it like a spatial relation where we try to determine something at an intermediate point from those right and left.

Again at the endpoint final endpoint $x_{n+1} = 0$, as there is nothing beyond the time horizon, so the derivative reads

$$\frac{\partial V}{\partial q_t} = \frac{\left(\sqrt{\frac{1}{bq_{t-1}}} - 1\right)}{q_{t-1}\left(\frac{q_t}{q_{t-1}} + \sqrt{\frac{1}{bq_{t-1}}} - 1\right)^2} - c$$

where the second term vanishes. For reference also record the first period condition

$$\frac{\partial V}{\partial q_t} = \frac{Q_1\left(\sqrt{\frac{1}{bq_{t-1}}} - 1\right)}{(q_1 + Q_1)^2} - \frac{\frac{q_2}{q_1}\left(\frac{1}{2}\sqrt{\frac{1}{bq_1}} - 1\right)}{q_t\left(\frac{q_2}{q_1} + \sqrt{\frac{1}{bq_1}} - 1\right)^2} - c$$

To get some intuition, the issue is like that of a boundary value problem with one fixed and another free endpoint.

Trying to find an algorithm that solves the problem would be most rewarding. Further, we might want to explain why the traditional Stackelberg conditions for leader

$$q_t = \frac{C}{4c^2}$$

and follower

$$Q_t = \frac{2c - C}{4c^2}$$

do not seem to fit any of the conditions above.

We may also consider an n-periodic solution, where $Q_n = Q_1, q_n = q_1$, infinite horizon where $n \to \infty$. If so, one may be concerned with the fact that the series does not converge. To remedy this, the natural choice by an economist would be to introduce discounting so that the profit series becomes

$$V = \sum_{t=1}^{t\to\infty} \left(\frac{q_t}{q_t + \sqrt{\frac{q_{t-1}}{C}} - q_{t-1}} - cq_t \right) \frac{1}{(1-r)^t}$$

Of course the interest $0 < r < 1$.

Reference

Puu T (2010) Dynamics of Stackelberg duopoly. In: Puu T, Panchuk A (eds) Nonlinear economic dynamics. Nova Science Publishers Inc., New York, pp 121–134

Chapter 15
CAUDEX QUINTUS

Abstract The Bertrand model for duopoly with close substitutes suggested in Chap. 5 was based on Kelvin Lancaster's restructured demand theory where not the marketed commodities, but their (metric) performance scores were entered in the utility functions. The marketed commodities were hence considered as bundles of such properties, designed by the producers. Lancaster assumed a simple linear structure for the connection between commodities and their properties. This way of seeing things provides a unique possibility to get a working definition of close substitutes, and to get a solid basis for modelling Bertrand style oligopoly. As we saw, pricing of substitutes with given design, could provide for some complex dynamics. However, the approach admits another perspective, i.e., not only competition through pricing, but also through product design. The Lancaster approach admits the calculation of shadow prices for the properties. The competitors can thus at each stage calculate the value of their current design, and they may compare it with the optimal design in each market situation, i.e., the design that given the technical production opportunities represents maximum value of the product. If this indicates that their product design is outmoded, they may go for a change. Redesign is costly and some conservatism is called for before a competitor goes for it. However, design provides for an alternative means of competition, and it might be interesting to explore different scenarios. Do the competitors converge on an identical design, or rather tend to different designs? Convergence on periodicity opens an even more interesting perspective; recurrence in design, "fashion cycles" to use a popular term from quite different kinds of study.

15.1 Product Design

This model was described in great detail in Chap. 5, where we only studied the Bertrand oligopoly case when the designs of the computing products was given.

Then, as the reaction functions could not be derived, we approximated them using combinations of square root and linear terms. The amplitudes A_1, A_2 were

T. Puu, *Disequilibrium Economics*, https://doi.org/10.1007/978-3-319-74415-5_15

determined by the design coefficients, and the iteration in adaptive form read

$$q_1' = (1-\delta)\, q_1 + \delta \left(\sqrt{A_1} \sqrt{\frac{q_2}{c_1}} - A_1 q_2 \right)$$

$$q_1' = (1-\delta)\, q_1 + \delta \left(\sqrt{A_2} \sqrt{\frac{q_2}{c_1}} - A_1 q_2 \right)$$

As we saw periodic and possibly even chaotic orbits could result.

We then tried to incorporate design changes in the model, assuming a rule that design would be changed to the optimal whenever the imputed value of the optimal design surpassed that of the actual design by a certain multiplier. Fortunately we then found that this first switching to optimal design made it possible to represent the design coefficients for the two properties as cosine and sine of one single angular coordinate θ_1, θ_2 for each competitor.

The calculated optimal design angle in any market situation was given by

$$\theta' = \arctan \frac{\lambda_2}{\lambda_1}$$

where λ_1, λ_2 were the current imputed property prices.

Then the competitors switched to optimal design according to the map

$$\theta_1' = \begin{cases} \theta_1 & \lambda_1 \cos\theta^* + \lambda_2 \sin\theta^* \leq k\,(\lambda_1 \cos\theta_1 + \lambda_2 \sin\theta_1) \\ \theta & \lambda_1 \cos\theta + \lambda_2 \sin\theta > k\,(\lambda_1 \cos\theta_1 + \lambda_2 \sin\theta_1) \end{cases}$$

$$\theta_2' = \begin{cases} \theta_2 & \lambda_1 \cos\theta + \lambda_2 \sin\theta \leq k\,(\lambda_1 \cos\theta_2 + \lambda_2 \sin\theta_2) \\ \theta & \lambda_1 \cos\theta + \lambda_2 \sin\theta > k\,(\lambda_1 \cos\theta_2 + \lambda_2 \sin\theta_2) \end{cases}$$

according to the second clauses, or otherwise kept previous design.

The amplitudes of the supply quantity dynamics reaction functions could then be calculated as

$$A_1 = \frac{\sin(2\theta_1)}{\sin(\theta_1 + \theta_2)}$$

$$A_2 = \frac{\sin(2\theta_1)}{\sin(\theta_1 + \theta_2)}$$

Finally, a simple updating rule was found for the imputed property prices.

$$\lambda_1' = \frac{1}{2} \frac{1}{q_1 \cos\theta_1 + q_2 \cos\theta_2}$$

$$\lambda_2' = \frac{1}{2} \frac{1}{q_1 \sin\theta_1 + q_2 \sin\theta_2}$$

This completes the supply-design map $q_1, q_2, \lambda_1, \lambda_2 \mapsto q_1', q_2', \lambda_1', \lambda_2'$ when the competitors use both supply and product design as means of competition.

The question is whether the competitors converge at the same design or choose different designs in case a fixed point is approached. Further it would be interesting to check if periodic solutions in design occur, producing fashion cycles, or if even more complex dynamics may occur.

Chapter 16
CAUDEX SEXTUS

Abstract In Chap. 7, we mentioned the migration model presented by Harold Hotelling in an MA thesis 1921, which has been completely forgotten, and apparently only been commented by two economists, the present author and a very close senior collaborator, professor Martin J. Beckmann. As a non-printed thesis it was not even available until Sir Alan Wilson published this very original work in Environment and Planning in 1978, but that too is now long ago and seems to have made no difference to its attraction. The model was reinvented twice, in genetics, and in mathematical ecology. Especially the latter, due to Skellam, 30 years after Hotelling, became a great success. The model combines Verhulst population growth with diffusion in space, based on migration from crowded areas. To judge from its title "A Mathematical Theory of Migration" it would seem to be highly useful for modelling today's population movements, but modelling follows fashion, and is, unfortunately, seldom inspired by issues in society. Anyhow, the model is, like the Hotelling duopoly discussed at length, set in one dimension, and Hotelling only provided two approximate solutions—close to zero population, and close to saturation population. We again want to take it to 2D geographical space. The original model is set in continuous space and time. In order to make it conform to the format of the present book, we suggest to discretize space to a square lattice, and time into periods. Growth alone, modelled by a logistic, might cause interesting new phenomena. As we know logistic growth in continuous time can be solved in closed form, whereas its discrete version became the most studied chaos model.

16.1 Migration

Hotelling's original aim was to explain the population movement westward in the United States. Despite the negligible interest from economists the model was extremely clever, and reinvented twice in other sciences: In genetics for modelling the diffusion of advantageous genes, and in ecology to explain growth and random dispersal of animal population. In particular the latter contribution by J.G. Skellam, 30 years after Hotelling, became truly seminal in ecology. One might wonder why nobody among contemporary economists seems to have become interested now when migration became a political issue of supreme concern.

T. Puu, *Disequilibrium Economics*, https://doi.org/10.1007/978-3-319-74415-5_16

We did not present the model in the main text, but this is easily remedied as it is simply stated and immediately speaks to intuition. The process Hotelling has in mind is stated in continuous time and space, but it is easily discretized to comply with the general format of this book.

After Thomas Malthus's classical population theory, population P is assumed to grow at a logistic rate

$$\frac{\partial}{\partial t}P = (1 - P)\,P$$

From very small populations it grows at an exponential rate, but as the sustainable population, here normalized to unity, is approached, Malthus's physical and moral checks come into play, and the process goes asymptotically to the sustainable value. Also, should initial population overshoot the sustainable, for instance due to a harvest failure, then the sustainable value is approached from above, people simply die of starvation.

Two comments are appropriate at this place: first, the sustainable population is taken as a given datum, which in the case of humans is not quite true as they produce their own means of subsistence, though eventually constrained by what is available on the planet. This can be changed easily by introducing normal production technology from economics, which the present author attempted several times (Puu 1985, 1991). We might also introduce current environmentalist issues such as adverse effects due to pollution. Second, the logistic process in continuous time is most well-behaved, even yielding to closed form first integrals, whereas in discrete time it became the basis for pedagogical chaotic processes. So, we should be aware that discretizing time introduces an element of instability, which was not there in the original model.

Further, as resources are given, or, alternatively, produced under decreasing returns, living standard per capita becomes lower the denser the population in a location is. As people are free to move and always did move, the would migrate from densely to sparsely populated regions.

Hotelling models spatial density differences by the Laplacian operator

$$\nabla^2 P = \frac{\partial^2 P}{\partial x^2} + \frac{\partial^2 P}{\partial y^2}$$

where x, y denote the space coordinates of geographical 2-dimensional space. The Laplacian, which is familiar from, for instance, heat diffusion measures conditions in a point with conditions in the immediate neighbourhood, all directions of departure taken in account. If $\nabla^2 P > 0$, then we have a sparsely populated location, and humans would choose to move in from the surroundings, if $\nabla^2 P < 0$ it is densely populated and humans would tend to move out. By the way, Skellam's argument was in terms of random movement. From local concentrations, more animals moving at random move out than in.

So taking both growth and migration in account, we would amend the local growth equation to

$$\frac{\partial}{\partial t} P = (1 - P)\, P + \frac{\partial^2 P}{\partial x^2} + \frac{\partial^2 P}{\partial y^2}$$

This is Hotelling's model. It is a PDE, partial differential equation. It has no simple solution. Hotelling simplified first to one dimensional space, and then to cases either close to zero and close to saturation, where linear approximations were good.

The full PDE is rather hard. More recently similar models have been studied in chemistry under the heading of reaction-diffusion equations.

However, what we propose here is to study this model numerically in discretized time and space (in two dimensions).

For simplicity take a square chessboard region with $4n^2$ location squares, each identified by indices i, j. The modelling is now obvious: the Laplacian can be constructed as

$$\frac{1}{4} \left(P_{i-1,j} + P_{i+1,j} + P_{i,j-1} + P_{i,j+1} - 4P_{i,j} \right)$$

Each location is surrounded by four other locations, West, East, South, and North, and the approximation builds on neighbourhood effects. Notably there exist other approximations also taking account of the additional four cells Southwest etc., which only share corners with the central cell.

The model now reads

$$P'_{i,j} = P_{i,j} + \delta \left(1 - P_{i,j}\right) P_{i,j} + \frac{\epsilon}{4} \left(P_{i-1,j} + P_{i+1,j} + P_{i,j-1} + P_{i,j+1} - 4P_{i,j} \right)$$

where δ, ϵ are suitably small parameters for constructing efficient numerical procedures.

To any PDE, and thus also, discrete version of it, belong boundary conditions for $P_{i,j}$ when $i, j = \pm n$. These can be different, for instance, constant when the region is insulated, protected against immigration/emigration, or free at the other extreme. One can also introduce forced immigration along the boundaries to study the effects of it on the interior.

The reader notes the obvious similarity to the game of life, though a bit more complex due to logistic growth instead of just on/off state. However a warning is in place if one thinks of the attractive blinking clusters eventually moving along the chessboard. This is deceiving, not locally, but globally because the game of life chessboard is a topological equivalent of a region that is a torus. There never existed one in geography, and will probably never exist, unless someone builds a huge wheel-shaped space station with many local communities.

References

Hotelling H (1921) A mathematical theory of migration. MA thesis, University of Washington, republished in Environment and Planning A10:1223–1239

Puu T (1985) A simplified model of spatiotemporal population dynamics. Environ Plan A17:1263–1269

Puu T (1991) Hotelling's migration model revisited. Environ Plan A23:1209–1216

Skellam JG (1951) Random dispersal in theoretical populations. Biometrika 38:196–218

Chapter 17
CAUDEX SEPTIMUS

Abstract As a rule the exponents of a Cobb-Douglas utility function are taken as given parameters. With ordinal utility their absolute values have no importance, though their ratio affects the preferences for competing commodities. In this simple stub we let them be determined trough advertising, more precisely the share of advertising expenditures for any product in total publicity spending. Assuming given unit costs (per newspaper line advert, per television second, and the like), we actually retrieve our first Cournot model, whose solution we know. However, if we also include production costs the system becomes much tougher and this is what we propose here. Assuming zero production costs was almost tradition in oligopoly theory, for instance Cournot, Hotelling, Palander, Stackelberg, all used it for simplification. The popular reference to mineral water, bubbling for free from the ground is not altogether convincing as standard case for production. For this reason we try to avoid such simplification as we have the advantage of computer power which the classics did not.

17.1 Advertising

Whenever we used Cobb-Douglas utility in this text we took the exponents α_1, α_2 in $U = q_1^{\alpha_1} q_2^{\alpha_2}$ as given parameters. However, once one includes publicity and marketing it is perfectly in order to make the exponents depend on advertising efforts. A simple choice is to put $\alpha_1 = \frac{x_1}{x_1+x_2}, \alpha_2 = \frac{x_2}{x_1+x_2}$, where x_1, x_2 measure advertising efforts. Such efforts obviously carry costs, which we may denote r_1, r_2 per unit. The utility function thus becomes

$$U = q_1^{\frac{x_1}{x_1+x_2}} q_2^{\frac{x_2}{x_1+x_2}}$$

Note that preferences written in this way imply that it is only the relative advertising effort that affects preferences. If the competitors increase efforts to the same degree, nothing changes. It is possible to construct exponents in a way so that joint efforts increase the general desirability for goods. But it is useless as long as we only have two commodities, the entire budget is spent on these goods anyhow—with

T. Puu, *Disequilibrium Economics*, https://doi.org/10.1007/978-3-319-74415-5_17

more, we could construct groups such that joint advertising increases demand for all commodities in each group.

The utility function is maximized subject to a budget constraint

$$p_1q_1 + p_2q_2 = 1$$

where p_1, p_2 denote commodity prices. Maximizing constrained utility we get $q_i = \frac{x_i}{x_1+x_2}\frac{1}{p_i}$, or writing the more convenient other way,

$$p_1 = \frac{x_1}{x_1 + x_2}\frac{1}{q_1}$$

$$p_2 = \frac{x_2}{x_1 + x_2}\frac{1}{q_2}$$

The profits for the two competitors become

$$V_1 = \left(\frac{x_1}{x_1 + x_2}\frac{1}{q_1} - c_1\right) q_1 - r_1x_1$$

$$V_2 = \left(\frac{x_2}{x_1 + x_2}\frac{1}{q_2} - c_2\right) q_2 - r_2x_2$$

Obviously c_1, c_2 denote manufacturing costs in addition to the publicity costs r_1, r_2.

Optimizing with respect to q_1, x_1, q_2, x_2 respectively, we find

$$q_1 = 0$$

$$q_2 = 0$$

$$x_1 = \sqrt{\frac{x_2}{r_1}} - x_2$$

$$x_2 = \sqrt{\frac{x_1}{r_2}} - x_1$$

The last two equations are exact replicas of our starting Cournot model, though in advertising efforts in stead outputs as variables. The first two indicate that we are back at the general problem with isoelastic demand: The best choice is to produce nothing and sell this at infinite prices, the revenue is always unitary and production costs are avoided if nothing is produced.

To avoid this anomaly we can set $c_1 = c_2 = 0$, quite as in so many oligopoly models in history. Then outputs become indeterminate, which is unsatisfactory, though the model for advertising can be run as such nevertheless. Some years ago, the present author gave this to a graduate business student, however he has not seen if anything published came out of it. Due to the snag the model is no good, and the results are known.

A possibility is to change the utility function, say to

$$U = (q_1 + a_1)^{\frac{x_1}{x_1+x_2}} (q_2 + a_2)^{\frac{x_2}{x_1+x_2}}$$

Rescaling the variables, we can always get $a_1 = a_2 = 1$. Then the inverse demand functions read

$$p_1 = \frac{x_1}{x_1 + x_2} \frac{1}{1 + q_1}$$

$$p_2 = \frac{x_2}{x_1 + x_2} \frac{1}{1 + q_2}$$

and the optimum conditions

$$\frac{x_1}{x_1 + x_2} \frac{1}{\left(1 + q_1'\right)^2} = c_1$$

$$\frac{x_2}{\left(x_1' + x_2\right)^2} \frac{q_1}{1 + q_1} = r_1$$

$$\frac{x_1}{x_1 + x_2} \frac{1}{\left(1 + q_2'\right)^2} = c_2$$

$$\frac{x_1}{\left(x_1 + x_2'\right)^2} \frac{q_2}{1 + q_2} = r_1$$

It looks like this iteration might yield interesting results, though the author never tried this out.

Chapter 18
CAUDEX OCTAVUS

Abstract In 1952–1953 Martin Beckmann proposed one of the most ingenious models that ever appeared in economics; for interregional commodity trade and pricing. In his short first article the entire problem was set in form of two short partial differential equations. For comparison, Paul Samuelson the same year dealt with the same problem for a discrete set of locations, in a way which makes an extremely clumsy impression if one has ever seen Beckmann's model (and understood it, which has been difficult for economists unfamiliar with differential operators other than derivatives). The main idea is that there are local excess supplies and demands that have to be levelled through trade in a 2D space region, and that there is a given transport cost field reflecting local transport facilities. The solution is that local price reflects local scarcity and that trade flows in the direction of the price gradient as reflecting maximum earnings from transportation. Moreover, competition among transporters puts a pressure on profit margins so that in the flow direction price differences exactly equal transportation cost. Later, in 1977 Beckmann proposed a dynamic version of the model, for temporal price change and reorientation of trade flow. This provides the basis for the proposed discretized model in this stub. We consider two alternatives, the full model, which has a number of snags, and a simpler model involving just interregional price formation which can be distilled and separated from the complete model.

18.1 Interregional Trade

In this somewhat extensive stub, which was not discussed in the main text, we will deal with one of the most ingenious models that ever appeared in economics due to Martin J. Beckmann. It deals with the distribution of prices and commodity trade in extended geographical space, and was cast in the format of three partial differential equations in time, and the two geographical dimensions of natural space.

The first model appeared in 1952 and focused the equilibrium price distribution and trade pattern. In 1977 it was cast as a dynamical system.

Like Hotelling's migration model it was no success, probably due to similar reasons, the unwillingness of economists to learn vector analysis with its basic operators, gradient, divergence, and Laplacian. It is not harder than several tools

T. Puu, *Disequilibrium Economics*, https://doi.org/10.1007/978-3-319-74415-5_18

that economists use, and working in the two dimensions of geography simplifies things enormously in comparison to the normal setting of hydrodynamics in the three dimensions of classical physics. Indeed, economists who did not understand the essence of the model called it "hydrodynamic analogy", though there is no optimization in hydrodynamics, which is the main issue in Beckmann's model, even if the divergence operator is used to relate sources and sinks of flows to local excess supplies and excess demands.

Unlike other spatial issues we dealt with, there are no localized firms in Beckmann's model, they are smeared out everywhere in the region studied, like the consumers, and in combination with the latter create local excess supplies or demands. These are supposed to depend on local prices; there is local excess supply where price is high and excess demand where price is low. The local structures giving this result were not analyzed in the original work.[1]

Along with this system of local excesses of supply and demand, there is also given a transport cost field over the region. It is not constant, but varies locally with, for instance, density, that determines the need of taking long detours, quality of roads, and delay due to congestion.

The solution sought then is to find a flow of merchandise that evens out local excesses of supply and demand. The facts of local scarcity are signalled through local price, so the trade flow is oriented in directions of highest ascending price, while seeking minimum transport cost routes. The final price distribution then equalizes spatial price differences to accrued transportation costs along the routes of flow.

18.2 The Static Model

All this is cast in but two short partial differential equations. In the static version that not involves time:

$$\nabla \cdot \boldsymbol{\phi} + q\,(p) = 0$$

and

$$k \frac{\boldsymbol{\phi}}{|\boldsymbol{\phi}|} = \nabla p$$

[1]They were studied in Beckmann and Puu (1985).

Here $q(p)$ is local excess demand (supply if negative), dependent on price with $\frac{d}{dp}q(p) < 0$. The field $k(x, y)$ is local transportation cost as given in the location x_1, x_2 in two-space.[2]

Two fields are to be *determined* by the model; the scalar field $p(x_1, x_2)$ which represents the price distribution over the region, and the vector field $\boldsymbol{\phi} = \left[\phi_1(x, y), \phi_2(x, y)\right]$ that represents the flow of traded commodities. A vector field has two components, direction $\frac{\boldsymbol{\phi}}{|\boldsymbol{\phi}|} = (\cos\theta, \sin\theta)$ and volume $|\boldsymbol{\phi}| = \sqrt{\phi_1^2 + \phi_2^2}$. The first is the actual direction of trade flow on a map of the region, and the latter is the volume of merchandise in the flow.

Let us also record the two differential operators, the price gradient

$$\nabla p = \left[\frac{\partial}{\partial x}p(x, y), \frac{\partial}{\partial y}p(x, y)\right]$$

and the flow divergence

$$\nabla \cdot \boldsymbol{\phi} = \frac{\partial}{\partial x}\phi_1 + \frac{\partial}{\partial y}\phi_2$$

The first operator tells us which is the space direction of highest price increase and how much price increases in that direction. The second gives us the change of flow due to local sources and sinks.

This latter explanation immediately says what Beckmann's first equation is about, it says that whatever there is of local excess supply is absorbed in the flow and whatever there is of local excess demand is drained from it. It thus states a local equilibrium condition.

As for Beckmann's second equation, it is a vector equation, consisting of two PDEs. It tells something of direction, i.e., that the flow is directed to the highest spatial increase of price, and further that in this particular direction price increases with transportation cost, neither more, nor less.

To the model also belongs a boundary condition, without which any PDE is incomplete. It can state that the flow stagnates along the boundary, if the region is insulated, for instance due to high tariffs, or else how it is involved in trade with the exterior.

[2] In Beckmann's model transport cost can also depend on the direction of transit, which is relevant, for instance, if we have a distance metric different from the Euclidean, for instance a Manhattan metric.

18.3 The Dynamic Model

Now, this is Beckmann's static model; we need the dynamic disequilibrium version of it, which we must try to discretize in order to get a computation model in the spirit of this book.

For the dynamic version

$$\frac{\partial}{\partial t} p = \lambda \left(\nabla \cdot \boldsymbol{\phi} + q(p) \right)$$

where λ is a positive adjustment speed constant. The right hand side is no longer zero in disequilibrium. Price is thus driven up by the sum of local excess demand and the surge from the flow of trade.

As for the flow adjustment we get

$$\frac{\partial}{\partial t} \boldsymbol{\phi} = \mu \left(\nabla p - k \frac{\boldsymbol{\phi}}{|\boldsymbol{\phi}|} \right)$$

where μ is another adjustment speed. This disequilibrium adjustment equation in its action upon vectors has two consequences. Again it consist of two equations; literally it is an action on each of the flow components, horizontal $\phi_1(x, y)$, and vertical $\phi_2(x, y)$. But we can also consider it as working on reorientation of direction $(\cos\theta, \sin\theta)$, and change of volume $|\boldsymbol{\phi}|$, the first when the actual flow no longer follows the price gradient (for instance because prices change in the dynamic process), the second if traders can reap more profit through increasing trade whenever any spatial price difference exceeds the cost of transportation.

18.4 Discretizing

We now try to put the model in a setting of discrete space and time. Consider again a square covered by a chessboard pattern of $4n^2$ small cells, or locations x_i, y_j, or just $-n \le i, j \le n$. The scalar fields for price and transport cost can be indexed $p_{i,j}$ and $k_{i,j}$, and also flow volume $|\boldsymbol{\phi}|_{i,j} = V_{i,j}$ to avoid the awkward vector norm symbol.

Direction is now more tricky. In this discrete format it is natural to restrict directions and spatial differences to West, East, South, and North only. We can still use the $(\cos\theta, \sin\theta)$ for direction, provided we restrict θ to an integral multiple of $\frac{\pi}{2}$

$$\theta = (r-1)\frac{\pi}{2}, 1 \le r \le 4$$

Then $(\cos\theta, \sin\theta)$ renders the vectors $(1,0)$, $(0,1)$, $(-1,0)$, and $(0,-1)$, and we can, for instance write $p_{i+\cos\theta, j+\sin\theta}$ to indicate price in any of the four neighbour cells.[3]

The price gradient can be obtained in the following way: Let us for a moment switch notation to $p_r = p_{i+\cos\theta, j+\sin\theta}$. Then define

$$P_{i,j} = \max_{r=1}^{4} p_r$$

and

$$r^* \in \{1,2,3,4\} \,|\, p_r = p_{r^*}$$

Then the gradient direction vector is

$$\left(\cos r^* \frac{\pi}{2}, \sin r^* \frac{\pi}{2}\right)$$

and the approximated gradient itself

$$\nabla p \approx \left(P_{i,j} - p_{i,j}\right)\left(\cos r^* \frac{\pi}{2}, \sin r^* \frac{\pi}{2}\right)$$

In the disequilibrium state the actual flow is not aligned with the price gradient direction $\left(\cos r^* \frac{\pi}{2}, \sin r^* \frac{\pi}{2}\right)$, so we need a different direction symbol, say φ, for the actual flow. Yet it still has to be in one of the four possible directions, say $\varphi = (s-1)\frac{\pi}{2}$.

The flow adjustment can be divided in two steps. First, we have to choose the new direction

$$s' = r^*$$

Further, if in this new direction price increases by more than the transportation rate then flow volume V is increased, in the contrary case it is decreased. Formally,

$$\begin{cases} \text{if } p_{i+\cos r^* \frac{\pi}{2}, j+\sin r^* \frac{\pi}{2}} > p_{i,j} + k_{i,j} \text{ then } V' > V \\ \text{if } p_{i+\cos r^* \frac{\pi}{2}, j+\sin r^* \frac{\pi}{2}} < p_{i,j} + k_{i,j} \text{ then } V' < V \end{cases}$$

We need to choose a precise rule for the change of flow volume, but we wait a moment as also price adjustment needs such a rule.

[3] A piece of warning is in place here. On the computer π is not an exact number, even if we define it through, for instance, $4\arctan(1)$. Hence $\left(\cos\left((r-1)\frac{\pi}{2}\right), \sin\left((r-1)\frac{\pi}{2}\right)\right)$ does not return a pair of integers, except for the case $r = 1$. As we want to use the numbers for indices we must invent some more roundabout way for determining gradient directions.

Turning to the latter issue, price change depends on two items, the surge from the flow given by some discrete approximation of the flow divergence $\nabla \cdot \boldsymbol{\phi}$ and the excess demand $z(p)$.

In the discrete approximation the divergence is particularly easy to deal with. The actual flow with volume $V_{i,j}$ has the direction $\varphi = (s-1)\frac{\pi}{2}$, so from the location i, j it goes to location $i + \cos s\frac{\pi}{2}$, $j + \sin s\frac{\pi}{2}$. Accordingly

$$\nabla \cdot \boldsymbol{\varphi} \approx V_{i+\cos s\frac{\pi}{2}, j+\sin s\frac{\pi}{2}} - V_{i,j}$$

As for excess demand we need some precise formula. The present author in working with the continuous version used

$$z = \frac{1}{p} - p$$

but anything decreasing from positive to negative will do.

As for adjustment of price we can use the same rule as in the asset model above

$$p' = p \exp\left(\lambda\left(\nabla \cdot \boldsymbol{\varphi} + z\right)\right)$$

again because linear adjustment as in the original continuous model may easily make the system explode or go to negative values. The same idea can be applied to flow volume adjustment

$$V' = V \exp\left(\mu\left(p_{i+\cos r^*\frac{\pi}{2}, j+\sin r^*\frac{\pi}{2}} - p_{i,j} - k_{i,j}\right)\right)$$

Recall that for flow direction adjustment we have a separate rule that aligns flow to price gradient.

Adding some boundary condition makes the agenda complete. One can start with the simplest case of insulation, i.e., for $i, j = \pm n$, $V = 0$ and $p = 1$, which with the excess demand function suggested brings prices in equilibrium in the boundary cells. This should provide all bits and pieces needed for cooking up a computer program. It is difficult to predict the outcome, in continuous space and time the system is extremely well behaved, but one never knows with the discrete setting.

18.5 Price Dynamic Alone

The original model can actually be broken up in two parts: Price dynamics; and trade dynamics, which means that the first issue can be dealt with alone.

To see how this works, take Beckmann's second (nondynamized) equation $k\frac{\boldsymbol{\phi}}{|\boldsymbol{\phi}|} = \nabla p$. The left hand side is composed of the scalar transport cost function and the *unit* vector $\frac{\boldsymbol{\phi}}{|\boldsymbol{\phi}|} = (\cos\theta, \sin\theta)$, whereas the right hand side is the price gradient vector $\nabla p = \left(\frac{\partial p}{\partial x}, \frac{\partial p}{\partial y}\right)$. If we take squares of both sides, then the right hand side becomes

$$(\nabla p)^2 = \left(\frac{\partial p}{\partial x}\right)^2 + \left(\frac{\partial p}{\partial y}\right)^2$$

As for the left hand side, we get transport cost squared k^2, whereas the unit vector multiplies out to a unit scalar, as $\cos^2\theta + \sin^2\theta = 1$. Thus

$$\left(\frac{\partial p}{\partial x}\right)^2 + \left(\frac{\partial p}{\partial y}\right)^2 = k^2$$

The right hand side is a given datum for the model, so we have just one partial differential equation for price alone. The form of this equation even suggests a constructive graphical method to solve the equation: Suppose we draw a curve for constant prices (equivalent to a boundary condition) on a map of the region studied. Then take a common compass, set the point anywhere on the curve, adjust the radius to the reciprocal of transport cost $\frac{1}{k}$ in that point, and draw a circle. The interior of the disk then represents the area of points to which traded goods can be moved using no more than one unit of money. Repeat the procedure for as many nearby points as needed, but remember to adjust the radius as transport cost varies from point to point. Then draw an envelope touching this family of circles. This then is an approximation of the next (in direction of flow) curve for constant prices, one monetary unit higher than the one we started from, and so the entire region can be filled with constant price curves. This digression is useful as we understand that implicit in the process is a dual set of flow curves (directions), orthogonal to the price curves.[4]

In this form Beckmann's model was never dynamized by himself, so we have no obvious way for the dynamization. The above description, however, implies that we on our square lattice of finitely many locations, have two ingredients for updating in a map: The price differences have to move closer to transport cost, and the flow directions need updating from any new price distribution.

So, suppose again that we define

$$p^*_{i,j} = \max_{r=1}^{r=4} p_{i+\cos r\frac{\pi}{2}, j+\sin r\frac{\pi}{2}}$$

and

$$r^* \in \{1, 2, 3, 4\} \Big| p_{i,j} = p^*_{i,j}$$

[4]A programmer more expert than the present author should be able to invent some clever computational procedure based on this idea.

Then the price gradient direction is

$$\left(\cos\left(\left(r^* - 1\right)\frac{\pi}{2}\right), \sin\left(\left(r^* - 1\right)\frac{\pi}{2}\right)\right)$$

However, our recent digression implies that we also have an actual (different in disequilibrium) flow direction.

$$\left(\cos\left((s - 1)\frac{\pi}{2}\right), \sin\left((s - 1)\frac{\pi}{2}\right)\right)$$

Accordingly one updating map rule is

$$s' = r^*$$

The other is much simpler than above; we need not be concerned with flow divergence or excess demand. Price differences over space just need to move towards actual transport cost, so take

$$p'_{i,j} = p_{i,j} \exp\left(\lambda\left(p^*_{i,j} - p_{i,j} - k_{i,j}\right)\right)$$

and this is it.

18.6 Suggested Agenda

We need to settle boundary conditions. One possibility is just constant price on the boundary, which does not promise interesting results.

More interesting is to assume a discretized version for our lattice of a transport cost dependent on radius vector, say $k = \frac{1}{\sqrt{x^2+y^2}}$ to mimic an extremely congested, almost inaccessible centre that is never transgressed by any transport, and try a boundary condition corresponding to $p = x^2 + y^2$. As a variant try a shear flow $p = x^2 - y^2$ or $p = xy$ on the boundary. Note that as an alternative we can deal with boundary conditions using different shapes of the region—if we want a saddle shape boundary condition, we can choose a saddle shape of the region, bounded by $\pm x^2 \pm y^2 = 1$, and constancy on the boundary.

In Fig. 18.1 we see what the outcome can be like with a mildly congested centre and constant price along the boundary of the 20 by 20 cell chessboard. The initial price distribution was randomly generated. The picture shows the outcome after five iterations. Colour is used to show the price distribution at this stage of the process, and the gradient directions are given by the horizontal and vertical segments, where a colour code was used to tell whether the gradient increases right, left, up, or down. When a horizontal or vertical line changes colour the gradient direction is reversed,

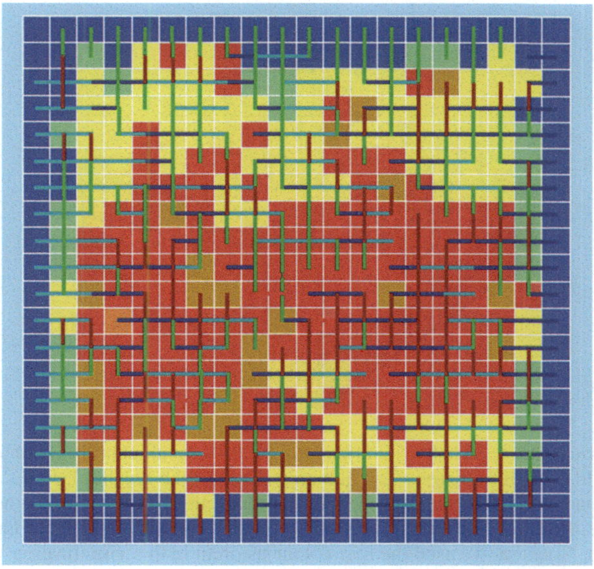

Fig. 18.1 A way to visualize the pricing and gradient patterns. Price is shown in terms of colour (blue to red)—this could become more informative if we change colours (though limited to 15 in the Quick Basic freeware) to intensity shades of the same colour—easy in any picture processing but time consuming if it is a simple program such as Paint. The gradient direction fields are, of course, horizontal/vertical, though they can have two directions each, so we used colour even for this. Dark blue right, grey left, green up, red down

if both the horizontal and vertical lines do this at the same cross, there is a local minimum, maximum, or saddle point.

Running the simulation but 50 iterations the price distribution seems to approach a stationary pattern. This is due to a low adaptation coefficient λ that was chosen. With higher values overshooting can occur, and probably result in more complex dynamics.

The colouring of the location cells (from blue to red) in Fig. 18.1 gives but a vague indication of the price, so we include also a Fig. 18.2 that shows the price surface after five iterations—the initial random price distribution is too irregular to be depicted as a surface. The surface is easy to calculate as far as the curves for constant i, j are concerned, but to see anything we need some work with any graphics program to colour the surface facets. It is good to remember the four colour theorem as this is what we need.

It would also be interesting to check (spatially) periodic boundary conditions in terms of trigonometric function with some wave number along each boundary segment, to see if this is reflected in the pattern of the interior.

It is possible that the price/direction pattern approaches a steady state, if so we can also use temporarily periodic boundary conditions.

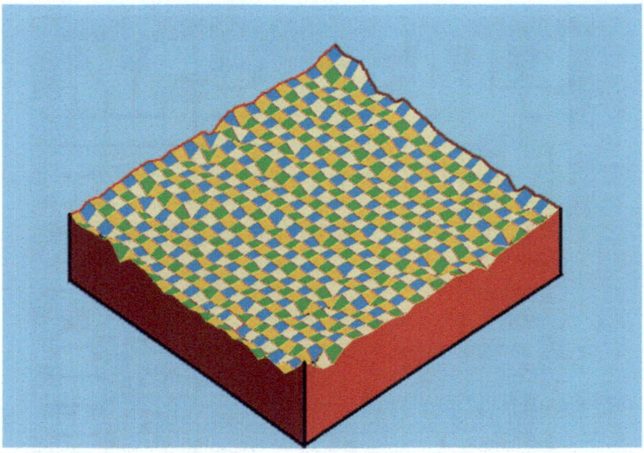

Fig. 18.2 The price surface after five iterations from a random initial price distribution. The initial pattern is too wild to be informative. The numerics gives us the height of points above locations in the bottom plane, and hence the lines connecting them; this can be done on the computer. For visualization one needs some picture processing program—in this case simple Paint was used to fill the rectangles. Note that we used the four colour theorem in this visualization

As for initial condition, we can, of course, in stead of the random, take a constant price over the entire square and find out if and how it breaks up. Another possibility is, again discretized, $p = x^3 - 3xy^2$, the manifestly unstable "monkey saddle" which is bound to break up under any circumstances.

A modern computer can be made to draw pictures of price surfaces at each stage of the evolution, or even a film of a sequence of such pictures.

Consider that computations with square lattices tend to be unstable, so triangulation may be a better choice. To comply with it we could make the region triangular or hexagonal.

Finally, it is worth recording that Beckmann's model works for other metrics than the Euclidean, for instance Manhattan. To deal with this in the continuous format is awkward, but in discrete space it can be done.

References

Beckmann MJ (1952) A continuous model of transportation. Econometrica 20:643–660

Beckmann MJ (1953) The partial equilibrium of a continuous space market. Welwirtschaftliches Arch 71:73–89

Beckmann MJ (1977) Equilibrium and stability in a continuous space market. Oper Res Verfahren 14:48–63

Beckmann MJ, Puu T (1985) Spatial economics: potential, density, and flow. North Holland Publishing Company, New York

Chapter 19
CAUDEX NONUS

Abstract Quite as in the case of migration the natural setting for business cycles is geographical space of two dimensions. Cycles are never but local phenomena, but spread through trade, import and export, from one location, region, or country to another. So, we can try to again put the model on square lattice, and let each cell have a multiplier-accelerator process of its own, though all the cells are now coupled through neighbourhood effects from adjacent cells. The coupling is easily obtained through a linear import/export propensity as once introduced by L.A. Metzler in 1950. As usual, a generalization in one sense must be compensated through some simplification of something else, as it is so easy to land on models that remain just statements that cannot be analyzed. Therefore we propose to sacrifice the present author's own model with capital, and to use a Goodwin type model where the floor and ceiling are both integrated in a single investment function with a cubic nonlinearity. There are many issues to study numerically. Like a partial differential equation such a discrete spatial model needs some boundary condition, which may stipulate insulation, trade with the exterior, or even periodic disturbances from the surrounding. Part of the boundary condition is the definition of the shape of the region. As neighbour cells may trigger a cell into prosperity or depression, it is natural if the "game of life" with its eventually moving and flickering flickering cell aggregates comes into mind. However, we must consider that the setting of this square is a topological equivalent to the surface of a torus, and nobody yet lived in a geographical region of torus shape. The author has seen some contributions in this direction, though this is sheer **abuse** of analogy.

19.1 Business Cycle Diffusion

As mentioned we do not include capital this time as both capital and income would make the model too complex. Instead, we incorporate not only floor, but also ceiling in the investment function, which takes a cubic shape. The model without the spatial setting was first proposed by the present author in 1989. It was worked later in 2004, where coauthor Dr Sushko added a clever analysis and much nice graphics.

T. Puu, *Disequilibrium Economics*, https://doi.org/10.1007/978-3-319-74415-5_19

To go straight to the point we change the linear accelerator to

$$I_t = a\left(Y_{t-1} - Y_{t-2}\right) - a\left(Y_{t-1} - Y_{t-2}\right)^3$$

The cubic term limits the action of the linear term with both a positive maximum (ceiling) for increasing income and a negative minimum (floor) for decreasing income. As always with polynomial approximations, the curve bends back, which at first sight may seem questionable. But, it is not quite as unrealistic as one may think. There are always investments in infrastructure and the like which need not be carried out at the moment. In booms resources are scarce and expensive, in slumps more abundant and cheap, so why not reduce investment programme in a boom and wait for the next slump? Further, most governments use investment programmes which on purpose act countercyclically. Of course, if we get too far on the wrong side from the extrema, it is no good, but it is a purpose for the total model construction to take care that this does not happen.

In one publication 2003, we actually replaced the cubic by horizontal floor and ceiling limits, and included interregional trade though only for two regions, which is useful for the spatial setting.

As for consumption, again take

$$C_t = cY_{t-1}$$

Combining, we get the reduced form iteration

$$Y_t = cY_{t-1} + a\left(Y_{t-1} - Y_{t-2}\right) - a\left(Y_{t-1} - Y_{t-2}\right)^3$$

For convenience define $Z_t = Y_{t-1}$. Then we can state the map

$$\begin{cases} Y' = cY + a\left(Y - Z\right) - a\left(Y - Z\right)^3 \\ Z' = Y \end{cases}$$

where we skipped the time index in order to provide space for other indices and used the dash for period advancement. The original model was more sophisticated than that, but we now want to keep the system as simple as possible, so we keep just this piece. The cubic iteration is good enough for creating interesting scenarios, including chaos.

What we now need is to put the model in the setting of an open economy with interregional trade, so suppose there is an exterior with the income variable X_t and a constant import export multiplier m. Then the income identity changes as we have to subtract imports mX_{t-1} and add exports mY_{t-1}.

$$Y_t = cY_{t-1} + a\left(Y_{t-1} - Y_{t-2}\right) - a\left(Y_{t-1} - Y_{t-2}\right)^3 - m\left(Y_{t-1} - X_{t-1}\right)$$

Using the map format, we have

$$\begin{cases} Y' = (c - m) Y + a (Y - Z) - a (Y - Z)^3 + mX \\ Z' = Y \end{cases}$$

As for X it represents income in the neighbourhood, and it is presumably governed by a similar system. In the spatial chessboard setting we again have four surrounding cells, so X represents the sum of incomes in all these cells. An economy that has a constant import propensity, which imports four times as much just because the surrounding cells are more numerous would soon run into an awkward balance of trade, so X should rather be treated as the average than the sum of incomes in the surrounding.

Discretizing we get the map

$$\begin{cases} Y'_{i,j} = (c - m) Y_{i,j} + a \left(Y_{i,j} - Z_{i,j}\right) - a \left(Y_{i,j} - Z_{i,j}\right)^3 + mX_{i,j} \\ Z'_{i,j} = Y_{i,j} \end{cases}$$

where

$$X_{i,j} = \frac{1}{4} \left(Y_{i+1,j} + Y_{i,j+1} + Y_{i-1,j} + Y_{i,j-1}\right)$$

Again we need boundary conditions, say as first approximation that $Y_{\pm n, \pm n} = 0$, i.e. there is complete rest at the boundary.

It is difficult to say what the solutions can be. The author knows something about similar systems in continuous format where the last trade expression is the Laplacian of income. For instance

$$Y'' = (a + c) Y' - aY'^3 + m \left(\frac{\partial^2 Y}{\partial x^2} + \frac{\partial^2 Y}{\partial y^2}\right)$$

has solutions with oscillating waves with plane sides (pyramids) proceeding at any of the three speeds that reduce the first order terms to zero. Whether these are attractive or not is difficult to say, though for similar models for overhead lines oscillating in strong wind, such saw-toothed waves have been proved to be attractive, though in one dimension nota bene.

References

Puu T (1989) Nonlinear economic dynamics. Springer lecture notes in economics and mathematical systems, vol 336. Springer, Berlin, pp 1–119

Puu T, Sushko I (2004) A business cycle model with cubic nonlinearity. Chaos Solitons Fractals 19:597–612

Sushko I, Puu T, Gardini L (2003) The Hicksian floor–roof model for two regions linked by interregional trade. Chaos Solitons Fractals 18:593–612

Chapter 20
CAUDEX DECIMUS

Abstract The 2 by 2 analysis of stock trade in an Edgeworth box in Chap. 8 was just a first formalization of a problem concerning multiplicity of equilibria, path-dependence and hysteresis in markets for durables that haunted the present author for decades. In this and the following stub, we try to relax some of the most restrictive assumptions. Here we first deal with the problem of more traders than two, as a matter of fact, an arbitrary number. No such rule for trade possibilities as could be stated for two traders is obvious, so we must add some more assumptions. Let us try to decompose the multilateral exchange problem in individual deals within buyer/seller pairs each of which we already know how to handle. Once a price is announced, we can split the set of traders into buyers and sellers, and within each group order them after how much they want to exchange. This ordering also reflects how much utility the traders can expect to gain. So we can assume that the most anxious buyers somehow find the most anxious sellers, as they have the strongest incentives. Once we consider a pair of traders, we know from Chap. 8 how to deal with their exchange. Of course, there remains an excess supply/demand within each pair after exchange, and eventually the list of pairs ends. There is no guarantee that the buyers sellers are equal in numbers, so this becomes another cause for remaining excess supply/demand. We are thus ready for the next round and the auctioneer announces a new price on the basis this excess. And so it goes.

20.1 Disequilibrium Trade: Several Agents

The analysis of trade with durable stock commodities in Chap. 8 was quite limited. As the issue is important though very little has been written about it, it would be most interesting to relax some of the most restrictive assumptions. We used the Walrasian idea of price dynamics dependent on excess demand or supply, but the Marshallian idea of quantity dynamics as dependent on the difference between demand and supply price may seem even more suitable for this analysis. This is at least how real auctions work.

Even more calling for generalization is the two by two model, chosen for convenience of illustrations in the Edgeworth box, to more commodities, which

T. Puu, *Disequilibrium Economics*, https://doi.org/10.1007/978-3-319-74415-5_20

brings in several relative prices and presumably more possibilities for interesting dynamics, and, above all, to more traders.

The step from two to three and more traders involves not only technical difficulties, but conceptual. We must make new rules for which potential traders will actually make transactions, who will be satisfied and who have to remain dissatisfied, and whether trade will be bilateral or multilateral in each step. As always in economic modelling we have to choose assumptions that are acceptable for intuition, but, weighing simplicity against realism, whatever we choose, our choice has to remain a bit arbitrary, and can reasonably be varied in major or minor detail. We will anyhow suggest one possible set of assumptions.

1. Assume an arbitrary number, say n, of potential traders whose wealth in terms of the two commodities as in Chap. 8 are denoted X_i, Y_i, $i = 1, \dots n$. Their preferences are given by Cobb-Douglas utility functions $U_i = x_i^{\alpha_i} y_i^{\beta_i}$, with $\alpha_i + \beta_i = 1$. Note that such "constant returns" make no sense with ordinal utility, it would work fine with $\alpha_i + \beta_i \neq 1$, but then we would have to rescale the coefficients to $\frac{\alpha_i}{\alpha_i+\beta_i}, \frac{\beta_i}{\alpha_i+\beta_i}$ when deriving the demand functions, which is unnecessary labour.
2. The budget constraints when wealth consists of quantities of the two commodities read $(x_i - X_i)+p\,(y_i - Y_i) = 0$, where p denotes the relative price, whereas the price of the first commodity as Walrasian "numéraire" is always unity.
3. Maximizing utility subject to budget constraints gives

$$\begin{cases} x_i = \alpha_i \left(X_i + pY_i\right) \\ y_i = \beta_i \left(X_i + pY_i\right) \end{cases}$$

4. We can now divide the potential traders in two groups, those that want to buy x and sell y, and those who want to do the reverse, so let us call the first group sellers and the last group buyers. For sellers $x_i > X_i$ or

$$\alpha_i \left(pY_i - X_i\right) > 0$$

for buyers $x_i < X_i$ or

$$\alpha_i \left(pY_i - X_i\right) < 0$$

We can now divide the traders in the two groups depending on the sign of $(pY_i - X_i)$, as the Cobb-Douglas exponent is always positive.

Note that the price is generated by some external agent trying to attain equilibrium, and that at each change of the announced price the grouping will be changed.

Further, suppose that in each group we order the traders after how much they will buy or sell. If we wish, we can introduce new indices si and bj within each group, where s alludes on seller and b on buyer. Now order the firms in each group according to how much they want to sell or buy respectively. So if lower numbers want to trade more we have

$$\alpha_{si}\left(pY_{si} - X_{si}\right) > \alpha_{si+1}\left(pY_{si+1} - X_{si+1}\right)$$

and

$$\alpha_{bi}\left(pY_{bi} - X_{bi}\right) < \alpha_{bi+1}\left(pY_{bi+1} - X_{bi+1}\right)$$

This is just definitions, and hence noncontroversial. Now we get to the more arbitrary assumptions.

5. Assume that somehow the greatest potential buyers, also those that have most utility to gain from exchange somehow find the most anxious sellers. This is not unnatural as both those are most keen to find an exchange partner.
6. Further, assume that each pair of traders thus makes a bilateral exchange just as described in Chap. 8. We will not repeat the details here as everything is written in the chapter. Each such exchange is limited by the trader in the pair who wants to exchange less, so there as a rule remains an excess demand or excess supply that in a latter stage may influence a new price. Also note that there is no guarantee that the sellers are exactly as numerous as the buyers, as a rule the numbers are different. Yet we assume trade to stop when the matching has exhausted any of the groups. This, of course, leaves a new source for excess supply or demand.

The present author is not certain that it is possible to make an elegant formal map with these assumptions, but the information is definitely sufficient for cooking up a computer program and make numerical explorations.

Chapter 21
CAUDEX UNDECIMUS

Abstract To continue the previous stub, we can also consider increasing the number of commodities. Supposing we still have two traders, what happens is that we need a solid 3D Edgeworth box. The only problems with this are visualization. What we mainly conclude from a 3D picture is that, as the budget constraint is a plane and not a line, we cannot say whether the optima are located on either side or the same side of the initial asset point. However, there are now two relative prices, and we can actually interpret one of the three commodities as numéraire. Then it becomes natural to consider partial deals of any of the two other commodities against the numéraire. It is just a matter of deciding which partial deal is the most advantageous, and to again reduce the problem to one deal of the kind we know how to deal with. At the end of the stub we suggest how to generalize the model to an arbitrary number of commodities and agents.

21.1 Disequilibrium Trade: Several Commodities

Another obvious generalization of the analysis of Chap. 8 is to consider more commodities. If the traders are still two, then we deal with a solid Edgeworth box with the traders represented in opposite corners. The size of the box is still determined by the total of commodities owned by both traders.

The indifference loci are now surfaces

$$U_i = x_i^{\alpha_i} y_i^{\beta_i} z_i^{\gamma_i}$$

and the budget constraint is a plane, fixed by two relative prices denoted p and r.

Denote initial commodity possessions for the first trader X, Y, Z in upper case as before. Normalizing the totals of the commodities owned by both traders to unity, the possession of the second trader is then $(1 - X)$, $(1 - Y)$, $(1 - Z)$. The budget constraint for the first trader is

$$x_1 + py_1 + rz_1 = X + pY + rZ$$

T. Puu, *Disequilibrium Economics*, https://doi.org/10.1007/978-3-319-74415-5_21

and for the second, when we measure x_2, y_2, z_2 in the same direction as x_1, y_1, z_1,

$$x_2 + py_2 + rz_2 = X + pY + rZ$$

If we consider indifference loci equivalent to the initial possessions, then

$$x_1^{\alpha_1} y_1^{\beta_1} z_1^{\gamma_1} = X^{\alpha_1} Y^{\beta_1} Z^{\gamma_1}$$

and

$$(1 - x_1)^{\alpha_2} (1 - y_1)^{\beta_2} (1 - z_1)^{\gamma_2} = (1 - X)^{\alpha_2} (1 - Y)^{\beta_2} (1 - Z)^{\gamma_2}$$

It is easy to calculate the optima for the two competitors

$$\begin{aligned} x_1 &= \alpha_1 (X + pY + rZ) \\ y_1 &= \frac{\beta_1}{p} (X + pY + rZ) \\ z_1 &= \frac{\gamma_1}{r} (X + pY + rZ) \end{aligned}$$

and

$$\begin{aligned} 1 - x_2 &= \alpha_2 ((1 - X) + p (1 - Y) + r (1 - Z)) \\ 1 - y_2 &= \frac{\beta_2}{p} ((1 - X) + p (1 - Y) + r (1 - Z)) \\ 1 - z_2 &= \frac{\gamma_2}{r} ((1 - X) + p (1 - Y) + r (1 - Z)) \end{aligned}$$

21.2 Some Geometry

Three dimensional issues are difficult to visualize without geometry, so we illustrate the facts in a solid Edgeworth box.

For this we need a numerical example, so take

$$\alpha_1 = \beta_1 = 0.25, \gamma_1 = 0.5$$

and

$$\alpha_2 = \gamma_2 = 0.25, \beta_2 = 0.5$$

Further set $p = 2, r = 4$, and $X = Y = Z = 0.5$.

The budget constraints are then

$$x_1 + 2y_1 + 4z_1 = \frac{7}{2}$$

and

$$x_2 + 2y_2 + 4z_2 = \frac{7}{2}$$

The indifference surfaces through the initial point are $x_1^{0.25} y_2^{0.25} z_2^{0.50} = \frac{1}{2}$ and $(1 - x_2)^{0.25} (1 - y_2)^{0.50} (1 - z_2)^{0.25} = \frac{1}{2}$ respectively. Note that these easily transform to more handy expressions

$$x_1 y_1 z_1^2 = \frac{1}{16}$$

$$(1 - x_2)(1 - y_2)^2 (1 - z_2) = \frac{1}{16}$$

The initial point before trade is the white dot in the middle of the box with all sides equal to unity, and the budget plane which passes through this point is drawn in yellow. The preference system is represented by two indifference surfaces that pass through the initial point, blue for the first agent, red for the second. Note the very small lens shaped area extending from the initial point containing points better than the initial for both competitors. The optimal points to which the competitors would like to move are also visible through the cutouts in the surfaces. See Fig. 21.1.

21.2.1 Digression on the Preparation of Solid Box Picture

The program for the 3D picture was drawn for a box with sides 2 by 2 by 2, where in the initial point the traders have one unit each of the commodities. This simplifies some formulas, but in essence there is no difference, a solid box with side 2 looks quite as one with side 1.

The intersection curves of the indifference surfaces with the planes

$$x, y, z = 1 \mp 1$$

are easily calculated from the formulas

$$xyz^2 = 1$$

$$(2 - x)(2 - y)^2 (2 - z) = 1$$

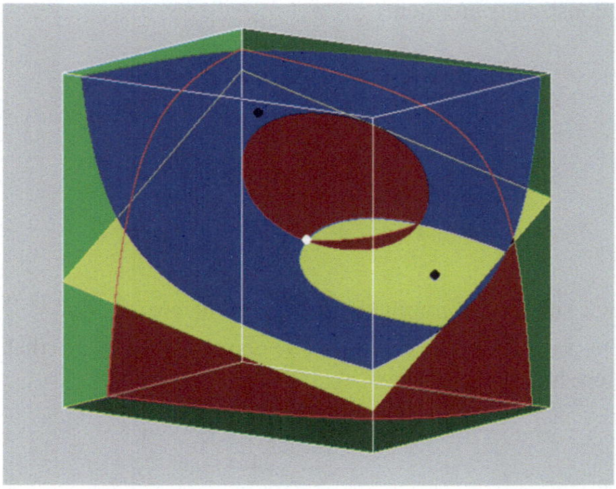

Fig. 21.1 In this 3D picture with commodities x_i, y_i, z_i along the axes (all sides of unit length), we take the initial point $X = Y = Z = 0.5$ and prices $p = 2, r = 4$. Accordingly, the budget plane is $x_1 + 2y_1 + 4z_1 = 3.5$, or, equivalently, $x_2 + 2y_2 + 4z_2 = 3.5$. The initial point is drawn in white and the budget plane in yellow. We also show the two indifference surfaces (blue and red respectively) that pass through the initial point, defined by $x_1^{0.25} y_2^{0.25} z_2^{0.50} = 0.5$ and $(1-x_2)^{0.25}(1-y_2)^{0.50}(1-z_2)^{0.25} = 0.5$. The Cobb-Douglas exponents are $\alpha_1 = \beta_1 = 0.25, \gamma_1 = 0.5$ for the first trader, and $\alpha_2 = \gamma_2 = 0.25, \beta_2 = 0.5$ for the second. We can easily calculate the desired optimum points which are shown in black. Compared to the two-commodity case it is difficult to say if they lie on "either side" of initial point or not. Anyhow, there is an intersection of the surfaces where the red one bulges up through the blue. It is the projection on the budget plane that matters, which is shown as the lens shaped cutout window. It is clear that *both* optimum points are far outside the yellow lens, so there seems to be little possibilities for trade. However, we propose to consider partial deals y for x or z for x, so reducing the situation to the familiar ones, as we see in the two following pictures of horizontal and vertical cuts through the solid Edgeworth box

and drawn on the box sides. Then any surface can be visualized by filling with colour in a picture processing program

More tricky are the intersection curves with the budget plane. We have to solve the systems

$$xyz^2 = 1, x + 2y + 4z = 7$$

and

$$(2-x)(2-y)^2(2-z), x + 2y + 4z = 7$$

respectively. The simplest way out is to solve the budget equation for x and substitute in the respective equations for the indifference loci. This results in quadratic equations for y (first trader)

$$y^2 + 2\left(z - \frac{7}{4}\right)y + \frac{1}{z^2} = 0$$

and z (second trader)

$$(2-z)^2 - 2\frac{3+2y}{8}(2-z) + \frac{1}{4(2-y)^2} = 0$$

The two real solutions of the quadratics give two sides of a closed curve (intersection of an indifference surface with the budget plane). As for x, it can be solved from the budget equation once we know y or z.

Finally we, as usual need to work out the plane projection $x, y, z \mapsto u, v$:

$$u = x \sin\theta - y \cos\theta$$

$$v = x \sin\phi \cos\theta + y \sin\phi \cos\theta + z \cos\theta$$

where θ, ϕ are the sight angles. As the present author does not access any research funds, he used QB45 which is freeware. There may be simpler to use software, but this logical old language gives a unique control of what is done. As for picture processing, there exist many simple drawing programs on the web, some even free. However, most of them are for handling photographs rather than technical drawings, so they always use some soft edge which blurs pictures, so the author used simple Paint which comes with Windows for the screen shots from QB45. It is though important to save all pictures in full resolution PNG, as otherwise the soft edge is there again.

Note how far outside the aforementioned small lens shaped area both optima are located. It is obvious that this is a situation where no trade at all is likely, so the prices must be much miscalculated for getting close to the equilibrium. Well, we might conceive of a tiny exchange within the lens shaped area, anywhere on a curve through the lens between points where indifference curves touch its sides. But the situation also raises an issue of principle.

21.3 Bilateral Deals

It is very hard to believe that two traders in three commodities find any point for multilateral exchange with all commodities. The fact that we have one commodity x_i as numéraire and two y_i and z_i with relative prices p and r suggest that we really treat the numéraire as a kind of money (even if it is included in the utility functions on its own), and so split the exchange in two separate bilateral exchanges x_i, y_i, with z_i fixed, and x_i, z_i, with y_i fixed. See Figs. 21.2 and 21.3 where the facts with partial exchange are illustrated on the horizontal and vertical planes. It is still true

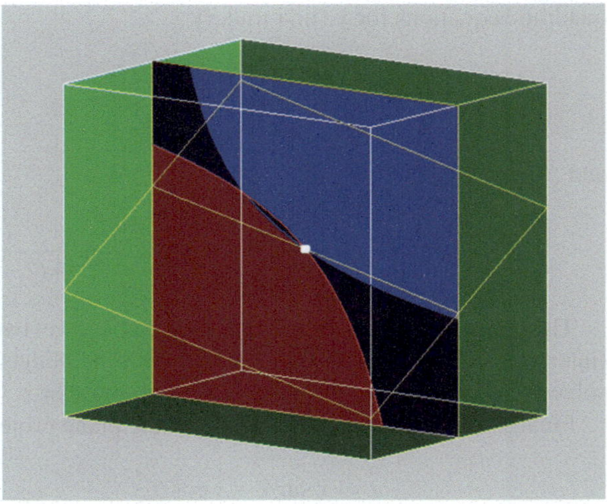

Fig. 21.2 A horizontal cut through the solid Edgeworth box showing possible partial deals x versus z. As we see, the lens for advantageous trade is very small, especially for trader 2 the initial point almost seems to be optimal, though things are difficult to see in this projection. The Figure is mainly included for illustrating the idea of partial deals

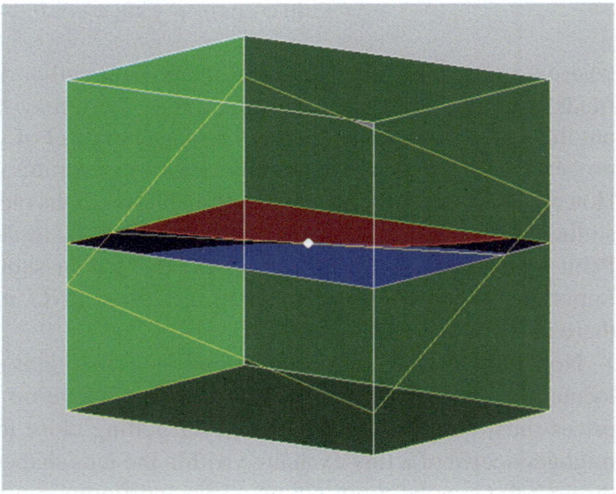

Fig. 21.3 In the vertical cut things show up more clearly. There is a small lens of advantageous trading x to z. But, if we consider tangency points of the indifference curves with parallels to the yellow budget line, they obviously fall on either sides of the initial point—a typical no trade situation. It is no better than in the 3D setting, though we now know how to classify the situation

that there might only be some tiny exchange, but the good thing is that we know how to handle this situation as we can directly jump into the formalism of Chap. 8.

The budget constraint for $z_i = Z = \frac{1}{2}$ fixed, now obviously reads

$$x_i + 2y_i = \frac{3}{2}$$

and the utilities to be maximized

$$U_1 = x_1 y_1$$
$$U_2 = (1 - x_2)(1 - y_2)^2$$

Likewise for $y_i = Y = \frac{1}{2}$ fixed

$$U_1 = x_1 z_1^2$$
$$U_2 = (1 - x_2)(1 - z_2)$$

and the budget constraint is

$$x_i + 4y_i = \frac{5}{2}$$

To calculate the optimum points is straightforward.

The cases are illustrated in two figures, in which the optimal points have not been drawn. Particularly the horizontal plane is not very distinct, but, never mind, we know how to deal with the problems in 2D and can jump right into the analysis of Chap. 8. See Figs. 21.2 and 21.3.

The utility functions work also with variable prices, but the budget constraints have to be rewritten

$$x_i + py_i = X + pY$$

and

$$x_i + rz_i = X + rZ$$

21.3.1 Bilateral Deal *x*, *y*

Accordingly we can immediately write the optima

$$x_1 = \frac{1}{2}(X_1 + pY)$$
$$y_1 = \frac{1}{2}\frac{1}{p}(X_1 + pY)$$

and

$$1 - x_2 = \frac{1}{3}((1 - X) + p(1 - Y))$$
$$1 - y_2 = \frac{2}{3}\frac{1}{p}((1 - X) + p(1 - Y))$$

21.3.2 *Bilateral Deal* $\mathbf{x}, z$

For the other bilateral deal the optima are

$$x_1 = \frac{1}{3}(X + rZ)$$

$$z_1 = \frac{2}{3}\frac{1}{r}(X + rZ)$$

and

$$x_2 = \frac{1}{2}((1 - X) + r(1 - Y))$$

$$z_2 = \frac{1}{2}\frac{1}{r}((1 - X) + r(1 - Y))$$

21.4 Trade Maps

We can immediately apply the map (8.8) repeated here

$$(X', Y') \begin{cases} (x_1, y_1) \text{ if } (y_1 - y_2)(y_1 - Y) \leq 0 \\ (x_1, y_1) \text{ if } (y_2 - y_1)(y_2 - Y) \leq 0 \\ (X, Y) \;\; \text{ if } (y_1 - y_2)(y_1 - Y) \geq 0 \end{cases}$$

where we just replaced the original condition written in terms of x_i, with one written in terms of y_i, the reason being that x_i appears in both partial maps and may cause confusion. The inequality signs are not affected as all points are on the downsloping budget line so all coordinate differences reverse sign by this switch.

We now also have another map for the partial deal x_i, z_i

$$(X', Z') \begin{cases} (x_1, z_1) \text{ if } (z_1 - y_2)(z_1 - Z) \leq 0 \\ (x_1, z_1) \text{ if } (z_2 - y_1)(z_2 - Z) \leq 0 \\ (X, Z) \;\; \text{ if } (z_1 - y_2)(z_1 - Z) \geq 0 \end{cases}$$

We must now resolve a little problem. Both maps may result in a change of X, as a rule different, so it is logical to assume that only one partial deal is made each period. The question is which.

There are several choices to make; the deals may, for instance, alternate, or we may make a random choice. This is good, though seems a bit mechanical. It seems better to incorporate an internal clause for the choice, for instance always choosing the deal which results in the largest change of the numéraire, as an indication of which deal is most useful at the moment, for both traders.

Thus the map is chosen for which $\left|X' - X\right|$ is maximal. For cases where no trade occurs the criterion value is, of course, zero.

Now, the price updating remains to be settled. Let us assume the updating rule used in Chap. 8, though now also including a map for r.

$$p' = p \exp\left(-\lambda \left(y_2 - y_1\right)\right)$$
$$r' = r \exp\left(-\mu \left(z_2 - z_1\right)\right)$$

Excess demands are as before, and also here we chose to formulate these in terms of excess of y_i, z_i which, of course reverses sign as compared to the original rule set in terms of x_i. Parameters λ, μ are positive adjustment speeds which determine how much overshooting there can be. Low values may produce smooth approach to equilibria, though very slowly, high values may result in more efficient approach, though they cause overshooting and more complex dynamics. Of course, clever variable adjustment speeds can also be tried.

For price updates there are several choices. We could choose to use only one price update, for the commodity for which partial trade occurs that period, or, perhaps, one period later. The simplest choice is to update both prices all periods, as things occur all the time through partial trade, and as tâtonnement is executed by an external third party.

We now have all bits and pieces for running a complete map after making the final choices of assumptions.

21.5 Arbitrary Numbers of Traders and Commodities

Note that splitting multilateral trade in bilateral deals also works with more commodities than three, of course, provided we skip ambitions at illustration. Just use the formalism for price adjustments (always number of commodities minus one) and the choice of maximally yielding exchange with the numéraire commodity.

The bilateral deals also give us a clue to how to generalize both to more agents and more commodities. Suppose we have m different commodities and, accordingly, $(m - 1)$ relative prices to the numéraire commodity, and n agents. Now for a start, allow only one transaction per period for each agent. Considering the possible bilateral deals as discussed in this stub, first establish which deal of a commodity against the numéraire the agent considers as most urgent. This can be established in terms of how much of the numéraire the agents wants to change. Once this is established for each agent they can be sorted in groups for commodities.

Then, within each commodity group we can proceed as suggested in the previous stub, i.e., order the buyers and sellers and pair them together. What remains is just to apply the main results of Chap. 8 to the bilateral deals. In this way the analysis can be generalized to arbitrary number of commodities and agents.

Chapter 22
CAUDEX DUODECIMUS

Abstract Interesting complex dynamics in oligopoly arises whenever the reaction functions have discontinuity points, as we saw in, for instance, the Cournot/Stackelberg case. There are many possibilities for this in more standard cases, arising either from the demand or from the cost function. A long article by Tord Palander from 1939 (unfortunately in Swedish) provides an entire inventory of such models. One of the simplest deals with competition between duopolists who have several production plants, maybe some originally suited for small scale production, easy to set in operation, but with steeply rising marginal costs, and some more modern suitable whenever demand calls for a larger load of production. Actually, such a firm has three alternatives, as it can also operate both plants dividing production load using the principle of equal marginal costs. As there are also fixed costs in any real situation, the three alternatives provide for a cost function with three segments and two jumps. It is easy to understand that combining two such reaction functions provide for interesting dynamic scenarios. Palander used linear functions, which involve more constraints for eliminating negativity of variables, so in this stub we propose to replace them with smooth demand and cost functions.

22.1 Duopolists with Multiple Plants

As the final stub we suggest the following problem. Assume two firms in Cournot duopoly which have different plants to operate, say an older plant appropriately used for small scale production involving low fixed costs, and a newer plant involving higher fixed costs but combined with more slowly increasing variable costs which is suitable when more has to be produced. This easily leads to discontinuous marginal cost functions, with resulting multiple optima in each situation and hence also discontinuous reaction functions and multiple Cournot equilibria. This model was suggested by Palander (1939) in his remarkable collection of intriguing duopoly models (regrettably only available in Swedish).

Two preliminary issues have to be dealt with. (1) Fixed costs do not necessarily lead to discontinuities if they cannot be avoided when the equipment is not in use; such are for instance costs for capital invested. However, in Chap. 2 we also discussed starting up costs incurred only when an equipment was prepared for being

T. Puu, *Disequilibrium Economics*, https://doi.org/10.1007/978-3-319-74415-5_22

taken in use. This is, as argued, a most relevant fact; it was only deleted from the agenda as it did not fit the concept of production functions. (2) With two plants each firm automatically has three options, the third resulting from using both plants and dividing the load according to microeconomic principles so as to equalize marginal costs. In such a combination marginal costs from combined operation increase very slowly, but the firm has to face the fixed costs for both plants.

So, using the CES function with built-in capacity constraints, and denoting production loads on the two machines q_1, q_2 for the first competitor and Q_1, Q_2 for the second, we have

$$q_i = \frac{k_i l_i}{k_i + l_i}$$

and

$$Q_i = \frac{K_i L_i}{K_i + L_i}$$

where the capital and labour inputs have obvious interpretations.

As we take capital k_i, K_i fixed we get production costs as usual

$$c_i = f_i + rk_i + w\frac{k_i q_i}{k_i - q_i}$$

$$C_i = F_i + rK_i + w\frac{K_i Q_i}{K_i - Q_i}$$

though we added the avoidable fixed costs f_i, F_i. Marginal costs obviously become

$$mc_i = w\frac{k_i^2}{(k_i - q_i)^2}$$

$$MC_i = w\frac{K_i^2}{(K_i - Q_i)^2}$$

so we can immediately get the optimal division of load between the plants through conditions

$$\frac{k_1}{k_1 - q_1} = \frac{k_2}{k_2 - q_2}$$

$$\frac{K_1}{K_1 - Q_1} = \frac{K_2}{K_2 - Q_2}$$

because we can just take the positive roots. Accordingly $k_2 q_1 = k_1 q_2$. and $K_2 Q_1 = K_1 Q_2$. If we put total output divided $q = q_1 + q_2$ and $Q = Q_1 + Q_2$ we can solve for

$$q_i = \frac{k_i}{k_1 + k_2} q$$

$$Q_i = \frac{K_i}{K_1 + K_2} Q$$

Accordingly the cost function for combined operation are

$$c_3 = (f_1 + f_2) + r(k_1 + k_2) + w\frac{k_1 + k_2}{k_1 + k_2 - q} q$$

$$C_3 = (F_1 + F_2) + r(K_1 + K_2) + w\frac{K_1 + K_2}{K_1 + K_2 - Q} Q$$

which we can also write

$$c_3 = f_3 + rk_3 + w\frac{k_3}{k_3 - q_3} q_3$$

$$C_3 = F_3 + rK_3 + w\frac{K_3}{K_3 - Q_3} Q_3$$

provided we define the fixed costs and capital stocks appropriately. Every item just sums up when we combine the two physical plants to a third virtual plant.

Marginal costs are now obvious, mc_i and MC_i still hold as stated above with $i = 1, 2, 3$.

As for revenue, we take our old friend the isoelastic case

$$p = \frac{1}{q + Q}$$

so, equating marginal revenue to marginal cost we have

$$\frac{Q}{(q + Q)^2} = w\frac{k_i^2}{(k_i - q)^2}$$

and

$$\frac{q}{(q + Q)^2} = w\frac{K_i^2}{(K_i - Q)^2}$$

where for each firm $i = 1, 2, 3$.

As usual, we can take the positive square roots and solve

$$q' = k_i \frac{\sqrt{\frac{Q}{w}} - Q}{k_i + \sqrt{\frac{Q}{w}}}$$

$$Q' = K_i \frac{\sqrt{\frac{q}{w}} - Q}{K_i + \sqrt{\frac{q}{w}}}$$

However, note that there are three possible solutions, local profit maxima for each firm, so the competitors have to decide which one to choose in each output interval. See Fig. 22.1. We assumed $i = 1$ for small outputs and $i = 2$ for larger, hence $i = 3$ renders the alternative for largest total output. Where the two break points are must

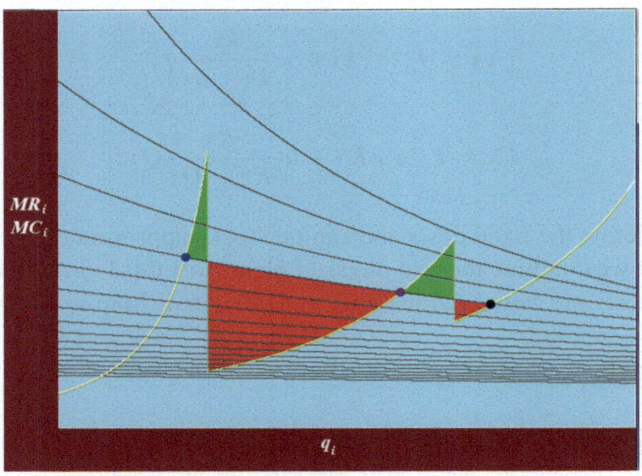

Fig. 22.1 Three local profit maxima for a marginal cost function with three rising sections. These are interrupted by vertical drops when the larger plant or eventually both are put in operation. We also see a family of marginal revenue curves. Economists usually compare the green gain areas to the red loss areas for judging whether a change of local optimum is better or not. Obviously the intermediate is worse than the leftmost, as the loss area is much larger. On the other hand the rightmost too is better than the intermediate, as at the corresponding move the green area is larger. Whenever the marginal revenue drops one step due to a move by the competitor this adds to the green areas and subtracts from the red, making a switch of local optimum profitable at some stage. We also understand that further moves make some local optima even disappear. At least this were so if there were no additional costs connected to putting the plants in operation, though now we have to consider also these in addition to the gain and loss areas. Yet the break points remain where they are in the picture

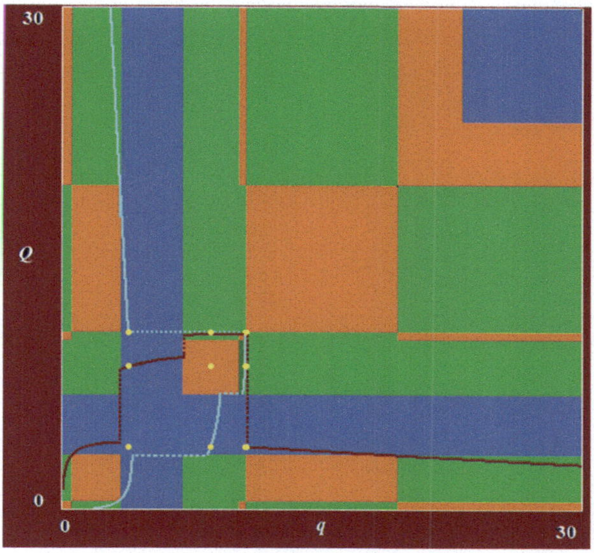

Fig. 22.2 Two identical firms with two plants each. Parameters: $k_1 = K_1 = 6.67$, $k_2 = K_2 = 22.5$ $f_1 = F_1 = 0.235$, $f_2 = F_2 = 0.1357$, $w = 0.01$. Each reaction function rises twice stepwise and then falls down to a very long downsloping section. There are three colours for different basins. The nine attractor points are divided between the different orbits

be established through considering equality for total costs

$$f_1 + rk_1 + w\frac{k_1 q}{k_1 - q} = f_2 + rk_2 + w\frac{k_2 q}{k_2 - q}$$

$$f_2 + rk_2 + w\frac{k_2 q}{k_2 - q} = f_3 + rk_3 + w\frac{k_3 q}{k_3 - q}$$

for the first competitor and similarly for the second. These equations boil down to simple quadratic equations, so we have to choose the roots that are appropriate. Finally, we have to establish where the composite three piece reaction functions hit the axes so as to take precautions to prevent the system from going to the origin. See Fig. 22.2.

The author tried his hands before with two mathematicians, Professor Laura Gardini and Dr Fabio Tramontana, though without taking such precautions.

There are different complications for those who like such, for instance cases with three plants for a firm; the number of operating combinations then increase from three to seven.

References

Palander TF (1939) Konkurrens och marknadsjämvikt vid duopol och oligopol. Ekonomisk Tidskrift 41:23–145, 222–250

Tramontana F, Gardini L, Puu T (2009) Cournot duopoly when the competitors operate multiple production plants. J Econ Dyn Control 33:250–265

The manufacturer's authorised representative in the EU is Springer Nature Customer Service Centre GmbH, Europaplatz 3, 69115 Heidelberg, Germany. If you have any concerns regarding our products, please contact ProductSafety@springernature.com

Printed and bound by CPI Group (UK) Ltd, Croydon, CR0 4YY
05/10/2026
02228122-0012